Geomorphology of Upland Peat

RGS-IBG Book Series

Geomorphology of Upland Peat
Erosion, Form and Landscape Change

Martin Evans and Jeff Warburton

WILEY-BLACKWELL

A John Wiley & Sons, Ltd., Publication

This edition first published 2010
© 2010 Martin Evans and Jeff Warburton

Edition history: Blackwell Publishing Ltd (hardback, 2007)

Blackwell Publishing was acquired by John Wiley & Sons in February 2007. Blackwell's publishing program has been merged with Wiley's global Scientific, Technical, and Medical business to form Wiley-Blackwell.

Registered Office
John Wiley & Sons Ltd, The Atrium, Southern Gate, Chichester, West Sussex, PO19 8SQ, United Kingdom

Editorial Offices
350 Main Street, Malden, MA 02148-5020, USA
9600 Garsington Road, Oxford, OX4 2DQ, UK
The Atrium, Southern Gate, Chichester, West Sussex, PO19 8SQ, UK

For details of our global editorial offices, for customer services, and for information about how to apply for permission to reuse the copyright material in this book please see our website at www.wiley.com/wiley-blackwell.

The right of Martin Evans and Jeff Warburton to be identified as the author of this work has been asserted in accordance with the UK Copyright, Designs and Patents Act 1988.

Wiley also publishes its books in a variety of electronic formats. Some content that appears in print may not be available in electronic books.

Designations used by companies to distinguish their products are often claimed as trademarks. All brand names and product names used in this book are trade names, service marks, trademarks or registered trademarks of their respective owners. The publisher is not associated with any product or vendor mentioned in this book. This publication is designed to provide accurate and authoritative information in regard to the subject matter covered. It is sold on the understanding that the publisher is not engaged in rendering professional services. If professional advice or other expert assistance is required, the services of a competent professional should be sought.

Library of Congress Cataloging-in-Publication Data

Evans, Martin (Martin Grant), 1970–
  Geomorphology of upland peat : erosion, form, and landscape change /
Martin Evans and Jeff Warburton.
    p. cm. – (RGS-IBG book series)
  Includes bibliographical references and index.
  ISBN-13: 978-1-4051-1507-0 (hardcover : alk. paper) ISBN 978-1-4443-3741-9
  1. Peatlands. 2. Peatland ecology. 3. Soil erosion. I. Warburton, J. (Jeff)
II. Title.

  GB622.E93 2007
  551.41–dc22

                                                                    2006032840

A catalogue record for this book is available from the British Library.

Set in 10/12pt Plantin by Toppan Best-set Premedia Limited
Printed and bound in Malaysia by Vivar Printing Sdn Bhd

1   2010

For Juliet, Daniel and Anna
and for
Christine, Isobel and Katie

Contents

Series Editors' Preface

The RGS-IBG Book Series only publishes work of the highest international standing. Its emphasis is on distinctive new developments in human and physical geography, although it is also open to contributions from cognate disciplines whose interests overlap with those of geographers. The Series places strong emphasis on theoretically informed and empirically strong texts. Reflecting the vibrant and diverse theoretical and empirical agendas that characterize the contemporary discipline, contributions are expected to inform, challenge and stimulate the reader. Overall, the RGS-IBG Book Series seeks to promote scholarly publications that leave an intellectual mark and change the way readers think about particular issues, methods or theories.

For details on how to submit a proposal please visit: www.rgsbookseries.com.

Kevin Ward
University of Manchester, UK

Joanna Bullard
Loughborough University, UK

RGS-IBG Book Series Editors

Acknowledgements

Much of the work reported in this volume has been the result of collaborations with postgraduate students. We are grateful for the shared experience of studying peatland erosion and their stimulating company. In particular, we would like to thank the following. In Durham: Richard Johnson, Andrew Mills, Duncan Wishart, Vicky Holliday, Alona Armstrong, Simon Foulds and Sarah Clement, and in Manchester: Juan Yang, Amer al-Roichdi, James Rothwell, Steve Daniels, Sarah Crowe, Laura Liddaman and Richard Pawson. Specifically we would like to thank Andy Mills for comments on Chapter 5. Collectively, their work has been an inspiration and has greatly contributed to many of the ideas in this book.

We would also like to acknowledge valuable discussions with colleagues: Tim Allott, Julia McMorrow, John Lindsay, Clive Agnew and Jeff Blackford (Manchester University), Tim Burt and Fred Worrall (Durham University), Alan Dykes, Joe Holden (University of Leeds), Louise Heathwaite (Lancaster University), John Adamson (CEH), Nick Haycock (Haycock Associates and Manchester University) and Penny Anderson (Penny Anderson Associates).

Thanks are due to English Nature for permission to work at the Moor House and Upper Teesdale Nature Reserve and to John Adamson for facilitating all our research at that site. In the Southern Pennines the National Trust and in particular Steve Trotter have been very supportive of our work. The support and assistance of the Moors for the Future partnership, and particularly Aletta Bonn and Cath Flitcroft have been an important part of the work on the peatlands of the Peak District.

We are grateful to the Departments of Geography at Manchester and Durham Universities for providing the resources to produce this book. The figures were prepared by Nick Scarle at Manchester and by the

Design and Imaging Unit in Durham, in particular Chris Orton. Both Nick and Chris have provided rapid, professional and patient responses to our multitude of requests. Our research has been supported both in the field and in the laboratory by John Moore and Mike Clarke (Manchester), Derek Coates, Alison Clark, Frank Davies, Eddie Million and Neil Tunstall (Durham), to whom goes our gratitude.

Funding for the research reported here has been provided at various times by Manchester University, Durham University, The Royal Geographical Society, The Royal Society, Moors for the Future, the British Geomorphological Research Group and NERC.

Upland peatlands can be inhospitable places, and the data which make up our understanding of these systems is hard won often under inclement conditions. We want therefore to acknowledge the passion and the persistence of the numerous peatland researchers past and present on whose work we have built here.

Finally the biggest thanks go to our families who have indulged what seemed at times to be an endless task.

Figure and Table Acknowledgements

The authors and publishers are grateful to the following for permission to reproduce copyright material.

Figure 1.1 is reproduced with permission from Charman, D. (2002), *Peatlands and environmental change*. Chichester: Wiley, 301p., © 2002, John Wiley and Sons Limited.

Figure 1.2 is reproduced with permission from Lindsay, R. (1995), *Bogs: The ecology, classification and conservation of ombrotrophic mires*. Edinburgh: Scottish Natural Heritage, 119p.

Figure 1.3 is redrawn with permission from an original diagram from O'Connell, C. (2002), Irish peatland conservation council blanket bogs information sheet. http://www.ipcc.ie/infoblanketbogfs.html. Accessed 6 October 2006.

Figure 1.4 is reproduced with permission from Lindsay, R., Charman, D., Everingham, F., O'Reilly, R., Palmer, M., Rowell, T. and Stroud, D. (1988), *The flow country: The peatlands of Caithness and Sutherland*. Peterborough: NCC.

Figure 1.6 is reproduced with permission of the author from Coupar, A., Immirzi, P. and Reid, E. (1997), 'The nature and extent of degradation in Scottish blanket mires'. In Tallis, J. Meade, R. and Hulme, P. (eds.) *Blanket peat degradation: Causes, consequences, challenges*. Aberdeen: British Ecological Society, pp. 90–100.

Figure 2.1 is reproduced with permission from Egglesman, R., Heathwaite, A. L., Grosse-Brauckmann, G., Kuster, G. E. Naucke, W. Schuch, M. and Schweikle, V. (1993), 'Physical processes and properties of mires'. In Heathwaite, A. L. and Gottlich, K. (eds.) *Mires, process, exploitation and conservation*. Chichester: Wiley, pp. 171–262. © 1993, John Wiley and Sons Limited.

Figure 2.2 is reproduced with permission from Price (2003), 'Role and character of seasonal peat soil deformation on the hydrology of undisturbed and cutover peatlands'. *Water Resources Research* 39(9): art. no.-1241. © 2003, American Geophysical Union.

Figure 2.3(a) is reproduced with permission from Fraser, C. J. D., Roulet, N. T. and Moore, T. R. (2001), 'Hydrology and dissolved organic carbon biogeochemistry in an ombrotrophic bog'. *Hydrological Processes* 15(16): 3151–66. © John Wiley and Sons Limited. Figure 2.3(b) is reprinted from Hoag, R. S. and Price, J. S. (1995), 'A field-scale, natural gradient solute transport experiment in peat at a Newfoundland blanket bog.' *Journal of Hydrology* 172(1–4): 171–84. © 1995, with permission from Elsevier.

Figure 2.4 is reprinted from Reeve, A. S., Siegel, D. I. and Glaser, P. H. (2000), 'Simulating vertical flow in large peatlands'. *Journal of Hydrology* 227(1–4): 207–17. © 2000, with permission from Elsevier.

Figure 2.5 is reproduced in part from Boatman, J. and Tomlinson, W. R. (1973), 'The Silver Flowe 1. Some structural and hydrological features of Brishie Bog and their bearing on pool formation'. *Journal of Ecology* 61: 653–66. © 1973, with permission from Blackwell Publishing Ltd.

Figure 2.7 is reproduced in part from Burt, T. P. (1992), 'The hydrology of headwater catchments'. In Calow, P. and Petts, G. E. (eds.), *The rivers handbook* (Volume I), Oxford: Blackwell, pp. 3–28. © 1992, with permission from Blackwell Publishing Ltd.

Figure 2.8 is reproduced from Burt, T. P. and Gardiner, A. T. (1984), 'Runoff and sediment production in a small peat covered catchment: Some preliminary results'. In Burt, T. P. and Walling, D. E. (eds.), *Catchment experiments in fluvial geomorphology.* Norwich: Geo Books, pp. 133–51. Reproduced with permission of the author.

Figures 2.7 and 2.9 reprinted from Evans, M. G., Burt, T. P., Holden, J. and Adamson, J. K. (1999), 'Runoff generation and water table fluctuations in blanket peat: Evidence from UK data spanning the dry summer of 1995'. *Journal of Hydrology* 221 (3–4): 141–60. © 1999, with permission from Elsevier.

Figure 2.10 is reproduced from Holden, J. and Burt, T. P. (2003), 'Hydrological studies on blanket peat: The significance of the acrotelm–catotelm model'. *Journal of Ecology* 91(1): 86–102. © 2003, with permission from Blackwell Publishing Ltd.

Figure 2.11 Reprinted from Worrall, F., Burt, T. P. and Adamson, J. (2004), 'Can climate change explain increases in DOC flux from upland peat catchments?' *Science of the Total Environment* 326(1–3): 95–112. © 2004, with permission from Elsevier.

peatlands'. *Canadian Journal of Soil Science* 82(1): 85–95. With permission from the Agricultural Institute of Canada and the authors.

Figure 6.6 and 6.7 are reprinted from Warburton, J. (2003), 'Wind splash erosion of bare peat on UK upland moorlands'. *Catena* 52, 191–207. © 2003, with permission from Elsevier.

Figure 6.9 is reprinted from Foulds, S. A. and Warburton, J. (2006), 'Significance of wind-driven rain (wind-splash) in the erosion of blanket peat'. *Geomorphology* In press. © 2006, with permission from Elsevier.

Figures 7.2, 7.3 and 7.4 are reproduced from Clement (2005) with permission of the author.

Figure 7.7 is reproduced with permission from Lindsay, R. (1995), *Bogs: The ecology, classification and conservation of ombrotrophic mires*. Edinburgh: Scottish Natural Heritage, 119p.

Figure 8.2 is reproduced with permission from Thorp, M. and Glanville, P. (2003), 'Mid-Holocene sub-blanket-peat alluvia and sediment sources in the Upper Liffey Valley, Co. Wicklow, Ireland'. *Earth Surface Processes and Landforms* 28(9): 1013–24. © John Wiley and Sons Limited.

Figure 8.3 is reproduced from Rochefort, L. (2000), '*Sphagnum* – a keystone genus in habitat restoration'. *The Bryologist* 103(3): 503–8. With permission of the author.

Figure 8.4 is reproduced with permission from Tim Burt and Sarah Clement.

Figure 8.6(a) is reproduced with permission from Sarah Crowe.

Figure 8.7(a) is reproduced from Skeffington, R., Wilson, E., Maltby, E., Immirzi, P. and Putwain, P. (1997), 'Acid deposition and blanket mire degradation and restoration'. In Tallis, J. H., Meade, R. and Hulme, P. D. (eds.), *Blanket mire degradation: Causes, consequences and challenges*. Aberdeen: Macaulay Land Use Research Institute, pp. 29–37. Reproduced with permission from the authors.

Figure 8.8 is reproduced with permission from McHugh, M., Harrod, T. and Morgan, R. (2002), 'The extent of soil erosion in Upland England and Wales'. *Earth Surface Processes and Landforms* 27(1): 99–107. © John Wiley and Sons Limited.

Figure 8.9 is reproduced with permission from Warburton, J., Evans, M. G. and Johnson, R. M. (2003), Discussion on 'The extent of soil erosion in upland England and Wales'. *Earth Surface Processes and Landforms* 28(2): 219–23. © John Wiley and Sons Limited.

Figure 9.3 is reproduced with permission from Rothwell, J. J., Robinson, S. G., Evans, M. G., Yang, J. and Allott, T. E. H. (2005), 'Heavy metal release by peat erosion in the Peak District, Southern Pennines,

UK'. *Hydrological Processes* 19(15): 2973–89. © John Wiley and Sons Limited.

Figure 9.4 is reproduced from Ehrenfeld, J. G. (2000), 'Defining the limits of restoration: The need for realistic goals'. *Restoration Ecology* 8: 2–9. © 2000, with permission from Blackwell Publishing Ltd.

Figure 9.5(a) is reprinted with permission from Dobson, A. P., Bradshaw, A. D. and Baker, A. J. M. (1997), 'Hopes for the future: Restoration ecology and conservation biology.' *Science* 277: 515–22. © 1997 AAAS.

Chapter One

Introduction

1.1 The Aims of this Volume

The aim of this monograph is to report recent work on the geomorphology of upland peatlands, and review current understanding of erosion processes and the long-term evolution of eroding upland systems. The book is written not only for peatland geomorphologists but also to provide a useful reference on current understanding of the physical functioning of peat landsystems for those working on their ecology, whether from a research perspective, or involved in practical management. In essence this book provides a state-of-the-art appraisal of understanding of the geomorphology of upland peats and demonstrates the importance of a geomorphological perspective for the understanding and management of these important and sensitive upland systems.

In this chapter we outline the scope of the book and provide a framework for evaluating the geomorphology of upland peat landsystems. First we consider the thematic and geographical context of the study. This is followed by explanation of some basic terminology and definitions used to describe peat and the classification of peatlands. We then discuss the geography of blanket mire complexes and examine patterns and causes of peat erosion. This is placed in the context of the evolution of peatland geomorphological science culminating in the development of a peat landsystem model which is used as a general framework for the book as a whole.

1.1.1 Thematic coverage

Upland peat is the residual product of the functioning of a series of fascinating and highly complex moorland ecosystems. As such it is hardly surprising that writing about peat landsystems has been dominated by biologists and ecologists (e.g. Gore 1983). Central to understanding these wetland systems has been an appreciation of their hydrology and there is an extensive body of literature describing the hydrological functioning of upland peatlands (see for example Ivanov 1981; Ingram 1983; Hughes and Heathwaite 1995b; Baird et al. 2004). However, in addition to their ecological functioning, upland peats are important terrestrial material stores. The slow continual accumulation of peat in intact peat bogs preserves a prehistoric archive interrogated by palaeo-ecologists and archaeologists alike (Charman 2002). When environmental conditions change, whether naturally or through human intervention, the continual accumulation of peat can be interrupted, and when the surface vegetation is stressed or removed the deep accumulations of organic sediment may begin to erode. Under these circumstances both the morphology and the ecological and hydrological functioning of the system becomes strongly influenced by erosion processes. This is an aspect of peatland functioning which has been relatively little studied.

This volume covers the hydrologically and ecologically controlled forms of intact upland mires but the majority of the book is concerned with the geomorphology of peatlands where processes of physical erosion are dominant. This focus is pertinent to mire management and conservation since it is in eroding peatlands where an understanding of their geomorphology is central to contemporary management and prediction of future mire condition.

1.1.2   Geographical context

The core of the book is focused on the authors' work on the eroding peatlands of northern Britain, particularly in the Pennine ranges, but every effort has been made to place this work in a wider context with reference to the most up-to-date work on upland mire systems. The United Kingdom (UK) has the most extensive erosion of upland peat in the world, and the vast majority of academic work on the causes, mechanisms and consequences of peat erosion is based on UK sites. Approximately 90 per cent of the published work on the geomorphology of upland peat refers to material derived from work in the British Isles. This fact, together with the geographical location of the authors' work, inevitably means that there is a strong UK focus to this book. However, the implications of what is reported extend beyond concerns with the management of erosion in the UK.

There is much debate over the causes of the extensive erosion in UK uplands but whilst severe land-use pressure has certainly been a factor, there is strong circumstantial evidence that climatic changes have played a major role. Increased storminess (Stevenson et al. 1992; Rhodes and Stevenson 1997) and desiccation of the mire surface (Tallis 1995) are both implicated and are effects which might be exacerbated across much of the world's northern peatlands under projected global climate changes (Houghton et al. 2001). UK peatlands are therefore an important field laboratory for the development of a thorough understanding of the dynamics of eroding peatlands. This will be essential in developing strategies to mitigate the possibility of enhanced physical degradation of wider northern peatlands in response to climate change, and the major effects on biodiversity, the carbon cycle and water quality which this would entail.

1.2   Terminology, Definitions and Peatland Geomorphology

There are several peatland classification schemes with terminology varying between nationalities and professional communities. Excellent summaries of the main classification types are given by Moore (1984) and Charman (2002). In this section the peatland terminology adopted in this volume is defined and the main types of upland peatland considered are identified.

1.2.1   Definitions of peat

Peat is an accumulation of the partly decomposed or undecomposed remains of plant material. There is a large range of peat types whose main properties vary depending primarily on the type of plant material composing the bulk of the organic matter and the degree of humification of the material. In most common soil classification schemes peat is usually treated as a distinct class. Even under specific organic soil classifications, peat is a distinct end member (Myślińska 2003). Under the widely accepted USDA Soil Taxonomy, organic soils form one of the main 12 soil orders and are known collectively as Histosols. Histosols contain at least 20–30 per cent organic matter by weight and are more than 0.4 metres thick. They have low bulk densities and high carbon contents. These soils occupy approximately 1.2 per cent of the ice-free land surface globally and are usually referred to as peats or mucks (McDaniel 2005). Peat deposits are also normally defined in terms of the depth of peat present in a particular setting but local definitions may vary. In a British and Irish context the criteria for separating peat from mineral soil varies. The Soil Survey of England and Wales uses 0.4 metres as the minimum depth for a peat deposit (Cruickshank and Tomlinson 1990), whilst 0.5 metres is

Table 1.1 Key properties of peat and examples of the importance of these for peatland geomorphic processes. This is not a comprehensive survey. It is included to demonstrate the important link between the peat material properties and controls on surficial processes

Peat Property	Importance	Reference
Basic Properties		
Water content of peat	The water content of peat can vary from about 200% to >2,000% of dry weight. The ability to store large volumes of water is the most striking characteristic of peat.	Hobbs 1986
Permeability and hydraulic conductivity	Permeability is a fundamental property controlling water movement and consolidation in peat. Permeability decreases markedly with depth with the abrupt transition from the acrotelm (aerated upper surface layers) to the denser catotelm (lower layers). Hydraulic conductivity may vary up to eight orders of magnitude between the layers.	Ingram 1983
Bulk density	The degree of decomposition and peat bulk density are intrinsically related. Decomposition decreases pore size. Bulk densities are low and variable.	Eggelsmann et al. 1993
Gas content	Gas content in peat may be as large as 5% of the volume. Most of this is free gas which influences permeability, consolidation and loaded pore pressures.	Hanrahan 1954
Organic (carbon) content	A high organic content is an intrinsic property of peat. Typically carbon contents of peat are approximately half the organic matter content which has important implications for terrestrial carbon stores.	Worrall et al. 2003
Micromorphology of peat	Important for water flow and rewetting in peat and secondary compression of the peat mass.	Mooney et al. 2000

Hydrogen ion activity and pH	Soil water pH is strongly correlated with vegetation and peat type and the chemistry of the water supply. Values range 3.5 to 6. Organic peat acid can be associated with weakened peat slopes	Söderblom 1974
Geotechnical Behaviour		
Geotechnical behaviour – standard index properties	Standard index (consistency) tests are not easily applied to peat material. Liquid limits are useful in characterizing certain types of peat but plasticity tests cannot easily be applied due to a lack of mineral clay.	Hobbs 1986 Carlsten 1993
Stress–strain – primary and secondary consolidation	By virtue of a very high water content peat is an extremely compressible material. Rapid consolidation is followed by secondary compression which is the dominant process.	Fox and Edil 1996
Changes in mechanical properties with organic content	No systematic relationship exists between mechanical properties and organic matter – soils behave in a complex manner due to differences in the amount and type of organic matter present.	Farrell et al. 1994
Flowing properties of peat slurry	Liquefaction of basal peat deposits, transport of material in a peat mass movement runout zone and transfer of organic material in river systems.	Luukkainen 1992
Peat creep	Slope instability and surface rupturing.	Carling 1986a
Shrinkage and desiccation	Peat is susceptible to shrinkage due to high water content. Desiccation cracking may promote delivery of surface water to the subsurface hydrological system promoting elevated pore pressures and peat mass failure.	Hendrick 1990
Thermal behaviour	Peat and other organic materials behave very differently in the cold: dry peat has a very low thermal conductivity due to high air content; wet saturated peat can have 5× higher thermal conductivity; whilst frozen peat 28× higher.	Seppälä, 2004

used in Scotland (Burton 1996), and 0.45 metres (undrained) in Ireland (Bord na Móna 2001).

1.2.2   The physical and geotechnical properties of peat

Consideration of the physical processes of erosion affecting peatland surfaces requires an understanding of the physical characteristics of peat as an earth material. Many of the challenges of a process-based approach to the geomorphology and hydrology of peatlands stem from the unusual properties of peat. Hobbs (1986) provides an excellent review of the properties and behaviour of peat. In this account Hobbs refers to peat as an 'ordinary extraordinary material' due to its unusual characteristics as an earth surface material (Table 1.1). For example, some properties of peat are similar to the behaviours of clay, but due to the extremely high water content of the peat, simple relations with material strength cannot be easily established (Landva et al. 1983). The material structure of peat greatly affects the hydraulic properties and strength of the deposit (Hobbs 1986). Although peat varies enormously, a 'typical' peat might be composed, by volume, of 85% water, 2% ash or mineral material, 8% organic material, and 5% air. The bulk density of the peat will increase as the organic matter becomes more decomposed but conversely the water content of peat decreases with decomposition. It is therefore essential to have a means of describing the different forms of peat so that the behaviour of these materials can be properly characterized.

Several peat description schemes have been developed. However, the von Post classification (von Post 1924) is widely used to provide a semi-quantitative description of the physical, chemical and structural properties of peat deposits. The scheme is based on semi-quantitative assessments of the principal plant remains, degree of humification, water content, fibre content and woody fragments. Hobbs (1986: 78–9) provides a succinct description of the main method. The von Post approach provides a rapid assessment method for characterizing peat properties. Table 1.1 considers examples of some of these key properties and identifies the important interrelationships between the basic peat components (phases) and the importance of these for the peatland geomorphic processes which are considered in more detail in subsequent chapters of this book.

1.2.3   Peatland classification

The term peatland is used, in this volume, to refer to all landscapes where the dominant surficial deposits are accumulations of organic matter (peat) in excess of 0.4 metres depth. The literature on the classification of peat landscapes is extensive but perhaps the most commonly adopted distinction is based on the source of water input to the peat mass. A distinction is made between *bogs* (ombrotrophic mires) and *fens* (minerotrophic mires) where the former are rainwater fed systems, typically acidic and nutrient poor, and the latter are groundwater fed, and typically circum-neutral with higher nutrient status (Hughes and Heathwaite 1995a). The term mire is used to refer to all forms of peatland, both bog and fen, and is the most appropriate term for many of the upland peatlands considered in this volume. Although typically dominated by ombrotrophic mire types these are complex upland systems with variable nutrient status. The definition adopted in this volume is closely related to the original definitions of mire and bog by Godwin (1941, 1956) which emphasize the nature of the sediments and the hydrological context and are appropriate to considerations of upland mires as geomorphological and hydrological systems.

The term 'upland' is widely used but needs clear definition in the context of this work. We conceive of the uplands as wildlands or areas where agriculture is extensive. As such we are working with a rather UK-specific definition of upland, akin to that of Ratcliffe (1977), of uplands as lands beyond the limit of enclosed cultivation. However, peatland landscapes of the type we are concerned with are not confined to the UK or indeed to a particular altitudinal band. They are often associated with resistant lithologies. The resultant thin soils tend to produce marginal lands, and indeed thin soils over impermeable bedrock tend to produce the waterlogged conditions favouring peat formation. Thus the low altitude peatlands of Newfoundland, Tasmania, the Shetlands and the Falkland Islands would fall within our definition of upland.

Hughes and Heathwaite (1995a) classify UK mires on a morphological basis into soligenous (sloping) mires, basin mires, valley mires, floodplain mires, raised mires and blanket mires. Charman (2002) suggests that this classification represents a generic hydro-morphological classification with broad applicability (Figure 1.1). The most widespread upland peat type is the blanket mire. Blanket bog is a term first defined by Tansley (1939) (Wheeler and Proctor 2000), to describe widespread ombrotrophic mire which follows the underlying topography like a blanket. Blanket bog is extensive and may therefore link other mire types into a continuous upland wetland system. A blanket bog may incorporate former basin mires in topographic low points and areas of raised mire formed either on summits, interfluves, or developed from former areas of basin mire. Where blanket peat is dissected or encompasses lines of pre-peat drainage then valley or floodplain mires form part of the complex, and where valley-side springlines are exposed soligenous mires may also

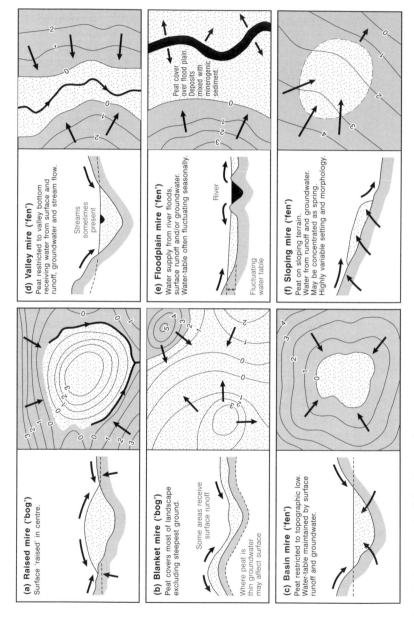

(a) Raised mire ('bog')
Surface 'raised' in centre.

(b) Blanket mire ('bog')
Peat covers most of landscape excluding steepest ground.

Some areas receive surface runoff

Where peat is thin groundwater may affect surface

(c) Basin mire ('fen')
Peat restricted to topographic low. Water-table maintained by surface runoff and groundwater.

(d) Valley mire ('fen')
Peat restricted to valley bottom receiving water from surface and runoff, groundwater and stream flow.

Streams sometimes present

(e) Floodplain mire ('fen')
Water supply from river floods, surface runoff and/or groundwater. Water-table often fluctuating seasonally.

Peat cover over flood plain. Deposits mixed with minerogenic sediment.

River

Fluctuating water table

(f) Sloping mire ('fen')
Peat on sloping terrain. Water from runoff and groundwater. May be concentrated as spring. Highly variable setting and morphology.

Figure 1.1   Hydro-topographical classification of mire types (after Charman 2002)

Table 1.2   Classification of scales of mire landforms (after Ivanov 1981)

Microtope	'A part of the mire where plant cover *and all other physical components of the environment connected with it* are uniform' Ivanov (1981: 6, emphasis added)
Mesotope	Isolated mire massifs with distinct patterns of microtopes and a single centre of peat formation
Macrotope	Complex mire massif formed from the fusion of isolated mesotopes through peat growth

form. Lindsay (1995) describes these mire assemblages, which span the full range of mire types identified by Hughes and Heathwaite (1995a), as blanket mire complexes. This usage is broadly synonymous with what we have termed upland mire complexes. Blanket peat is a necessary component of an extensive upland mire complex and in this volume, for reasons of style and respect for local usage, the terms upland mire complex, blanket mire complex and blanket peatland are used interchangeably.

Lindsay (1995: 22) notes that 'mire complexes are most frequently encountered in the uplands where several hydro-topographical units . . . fuse to form an extensive complex cloaking the landscape with peat.' To understand what Lindsay means by hydro-topographical units it is necessary to first review the basic classification of peatland landforms produced by Russian peatland scientists and summarized in Ivanov (1981) (Table 1.2). In the context of upland peatlands the macrotope is the upland mire complex identified above and the mesotopes which combine to form this macrotope might reasonably be characterized as any of the mire types identified by Hughes and Heathwaite (1995a). The mesotope therefore is essentially a unit at the scale of peat landforms and the macrotope describes the peatland landscape. In a sense the geomorphology of a landscape where peat formation occurs is the prime control over the mire type produced since local slope is the major determinant of the direction of groundwater flow relative to the centre of mire growth and consequently of the division between fens and bogs. Lindsay (1995) uses an explicitly geomorphological framework to subdivide elements of the blanket bog type following Ivanov's (1981) assertion that mire classification should include a geomorphological element. Lindsay describes watershed mires (probably better defined as summit mires), spur mires, saddle mires and valley side mires defined by their topographical setting. These hydro-topographical units are essentially mire mesotopes classified geomorphologically. Figure 1.2 illustrates the combination of a series

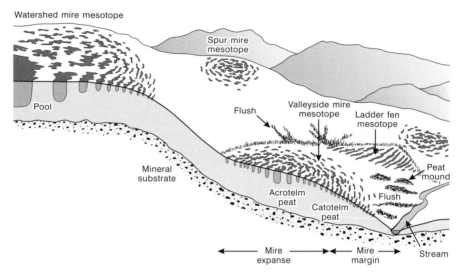

Figure 1.2   Combination of hydro-topographical units to form an upland mire complex (redrawn after Lindsay 1995)

of these hydro-topographical units to form an upland mire complex (macrotope).

At a smaller scale Lindsay et al. (1988) identify seven major microforms associated with UK blanket peatlands. These can be classified as follows:

<u>Hydro-ecological microforms</u>
  hummocks
  ridges, high or low
  hollows, *Sphagnum* or mud-bottomed
  pools, permanent or ephemeral
<u>Geomorphological microforms</u>
  erosion gullies
  erosion haggs
  peat mounds

The hydro-ecological microforms are largely controlled by the close interaction of hydrological and ecological processes on the mire surface (Belyea and Clymo 1998; Bragg 2002; Laine et al. 2004). These processes have formed the focus of the vast majority of previous academic work on upland peatlands. The geomorphological microforms and the physical processes which control them have received much less attention.

Therefore, in terms of the classification of upland peats outlined above the focus of this volume is the geomorphology of upland blanket mire complexes.

1.3   The Geography of Blanket Mire Complexes

As primarily ombrotrophic systems, the distribution of blanket mire is closely controlled by climate. All mire systems require a positive water balance for their long-term growth and maintenance. In ombrotrophic systems the key components of the water balance are precipitation inputs and losses by evapotranspiration (e.g. Evans et al. 1999; Kellner and Halldin 2002) (see Chapter 2). Positive water balance is favoured by higher rainfall, consistent with the observation that blanket peatland is the dominant peatland type in hyper-oceanic areas of the world. There are numerous statements in the literature regarding the threshold climate conditions for blanket bog formation. Pearsall (1950) suggested that in England a threshold precipitation of 1,250 mm existed. Lindsay et al. (1988) suggested a more realistic set of limiting conditions based on four key criteria: (1) annual precipitation above 1,000 mm; (2) >160 rain days per year; (3) warmest month with mean temperature <15°C; and (4) limited seasonal temperature variability.

The key controls on the nature of the local water balance are the relative rates of precipitation input and evaporative loss. Hence, the location of areas favourable for mire formation is also affected by parameters controlling evaporation such as temperature, relative humidity and wind speed. The inclusion in Lindsay's scheme of a measure of rainfall frequency relates to the requirement to maintain positive water balance and hence a high water-table despite evaporative losses which occur throughout the year. The geographical implication of considering the controls on evaporation is that the more oceanic the climate and the lower the mean temperature the lower the precipitation input required to maintain a positive water balance. Since mean temperatures decline with latitude and elevation this explains why on the Shetland Islands blanket bog occurs down to sea level whereas in the Southern Pennines of England it is confined to elevations above 500 metres despite the two locations having similar mean annual rainfall (circa 1,200 mm). A similar effect of altitude is demonstrated by the distribution of Irish bogs (Figure 1.3) where low elevation 'Atlantic Bogs' are distributed west of the 1,200 mm isohyet but 'Mountain Bogs' occur in all upland locations (O'Connell 2002).

Figure 1.4 (after Lindsay 1995) is a global map of the locations which fulfil Lindsay's (1995) criteria for the support of blanket peat. This is a

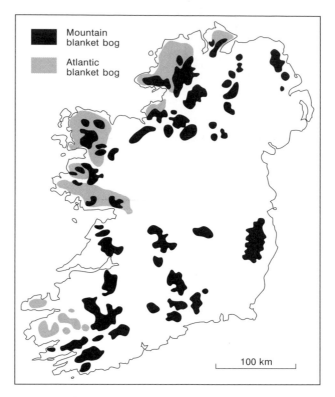

Figure 1.3   Distribution of blanket bog types in Ireland (redrawn after O'Connell 2002)

limited range of hyper-oceanic environments but blanket peat is recorded within all of these areas worldwide. In reality climatically-based predictions of the presence of blanket mire are simply a proxy measure using readily available data to approximate the water balance of a given region. It will be seen, however, in Chapter 2 that accurate measurement of the water balance in mires is far from straightforward so the climatic proxy approach is a useful initial approximation.

The primary research reported in this volume relates to the eroded upland blanket mire complexes of the UK. Where it exists we draw on a wider international literature relating not only to blanket mire but to all forms of ombrotrophic (rain fed) peatland. In part this is justified because of the range of mesotopes which may be encountered within a blanket mire complex, and in part it is recognition of the similarities of process across the spectrum of ombrotrophic peats.

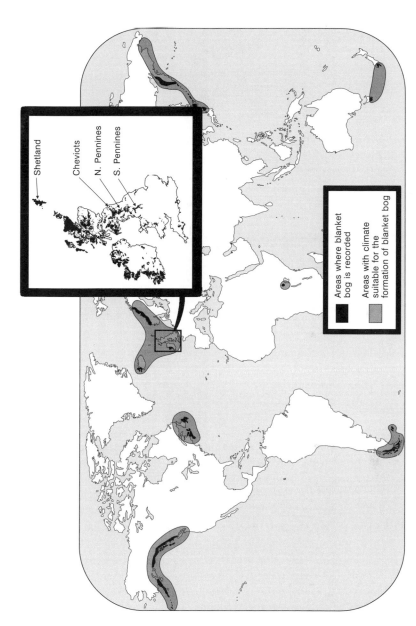

Figure 1.4 Global distribution of blanket bog (modified after Lindsay 1988)

Table 1.3 The extent of erosion of blanket peatlands in the UK and Ireland

Location	Estimated area of erosion	Notes	Reference
South Pennines, England	33 km² (6%) of eroded peat	Ground bare or partially bare	Philips et al. 1981
Moor House Nature Reserve, N. Pennines, England	8% eroded, 10% eroded and revegetated	Based on bare ground or morphological expression of gullying	Garnett and Adamson 1997
Wales	30% of peat degraded	Includes succession to less favourable mire types as well as physical erosion	Yeo 1997
Scotland	20% of blanket mire affected by gullying		Coupar et al. 1997
Scotland	6% of Scottish Uplands eroded 4.7% gullied		Grieve et al. 1994
Connemara, Ireland	27% of upland blanket mire eroding		Mckee and Skeffington 1997 (Geerling and Van Gestel 1997)
Wicklow Mountains, Ireland	33% of blanket mire affected by gullying 24% of blanket peat gullied		Cooper and Loftus 1998
Central and northwest Ireland			Large and Hamilton 1991
Northern Ireland	Blanket peats 29% eroded, 56% cut/drained, 15% intact	Total peat coverage estimated at 140,000ha	Cruickshank and Tomlinson 1988

1.4   Patterns of Peat Erosion in Space and Time

An important context for understanding the geomorphology of eroded peatlands is knowledge of the distribution of erosion in time and space. Severe and extensive erosion of upland peat is a phenomenon which is almost unique to the UK and Ireland. Outside this area Glaser and Janssens (1986) describe local peat erosion in Newfoundland, and Foster et al. (1988) describe minor natural erosion of bogs in Labrador and in central Sweden. Peat erosion through decay of palsa mires or in areas of thermokarst development has been reported from permafrost regions (e.g. Gurney 2001; Oksanen et al. 2001; Zuidhoff 2002). Short duration local peat erosion due to fire, or localized livestock impacts are common in many blanket mire systems. For example in the Australian Alps severe damage to upland *Sphagnum* bogs has been recorded due to severe fire in 2003 and subsequent trampling by livestock (Victoria National Park Association 2005). Pitkanen et al. (1999: 454) report that 'erosion has a negligible role in Finnish peatlands' and, although there are local exceptions associated with specific impacts on the mire surface vegetation (grazing, fire, peat mining, etc.), this statement holds true for most peatlands outside of the extreme western fringe of Europe.

Although work in Britain and Ireland has tended to emphasize the erosion of peat by running water, many of the reports of peat erosion from other regions of the world typically emphasize the importance of aeolian erosion (e.g. Luoto and Seppälä 2000 [Finnish Lapland]; Zuidhoff 2002 [northern Sweden]; Selkirk and Saffigna 1999 [sub-Antarctic Macquarie Island]). Wind erosion of milled peatlands has also received considerable attention in North America (e.g. Campbell et al. 2002; Lavoie et al. 2003), with the resultant instability of bare peat surfaces proving a significant impediment to attempts to restore mined peatlands. Recent work on aeolian erosion of peats is reported in Chapter 6 of this volume. In contrast to these findings, Klove (1998: 213) concluded 'that rain is the major cause of erosion from peat mine surfaces' in a study of sediment delivery from a peat mine in northern Finland. Uncertainty regarding the dominant cause of peat erosion was a major theme in the early literature on erosion in the UK, particularly the relative importance of wind and water (Bower 1961; Radley 1962).

In the UK and Ireland extensive peat erosion occurs across much of the blanket mire surface (Table 1.3, Figures 1.5 and 1.6). McHugh et al. (2002), in a survey of erosion across the uplands of England and Wales, demonstrated that peat soils in the uplands are the most severely eroded soil class. Overall the picture which emerges of the upland mires of the UK and Ireland is very different from the rather limited peatland erosion

Figure 1.5   Example of peat erosion from around the UK and Ireland (a) Severe erosion on the summit of Kinder Scout, South Pennines. (b) Peat slide scar at South Channerwick, Shetland. (c) Eroded and partially re-vegetated peat haggs on Hard Hill, North Pennines. (d) Peat slide scar at Doon Carton, Co. Mayo. (e) Severe gully erosion on the Bleaklow Plateau, South Pennines. (f) Sediment delivery to Cow Green reservoir via an eroding moorland grip system

reported from other parts of the world. Regionally extensive peat erosion in the UK is in strong contrast to the global picture of very local peat erosion associated with particular environmental impacts. Sheet and gully erosion of blanket mires is commonplace in the UK and Ireland and a major part of the surface patterning of many mires is controlled by geomorphological microforms.

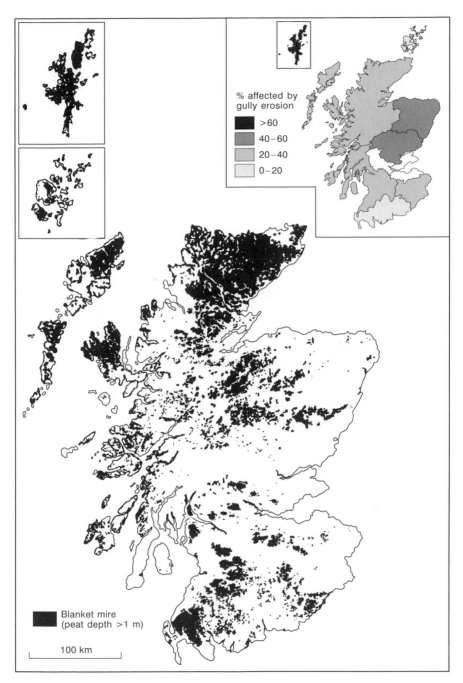

Figure 1.6   Distribution of gully erosion in Scottish blanket peats showing considerable regional variation in the extent of erosion (redrawn after Couper et al. 1997)

1.4.1   The onset of peat erosion

An explanation of the concentration of eroded upland mire in the UK and Ireland requires consideration of the reasons behind the onset of upland erosion. This question has preoccupied peat erosion researchers in the UK and Ireland for much of the last 40 years. The eroding upland blanket mires of the UK and Ireland support considerable depths of peat, typically in excess of a metre and locally up to 6 metres or more. It is clear therefore that the long-term character of these mires has been as sites of peat accumulation, but that at some stage in their history the nature of the mire system has switched from peat accumulation to erosion.

Arguably the most significant body of work on the initiation of peat erosion is that by Tallis (Tallis 1964a and b, 1965; Tallis and Switsur 1973; Tallis 1985a and b, 1987; Mackay and Tallis 1994; Tallis 1994; Tallis and Livett 1994; Tallis 1995, 1997a, b and c; Tallis et al. 1997; Tallis 1998). Tallis noted that the onset of significant gully erosion in a peat bog causes drainage and lowering of the water-table in intact peat immediately adjacent to the eroded gully. The vegetation and hydrology of blanket mire surfaces are closely linked so that the onset of erosion leads to changes in mire surface vegetation adjacent to the gullies. These vegetation changes are recorded in the peat stratigraphy as changes in the pollen and particularly plant macrofossil record. Gully-side peats represent an intact organic depositional sequence so they can be dated by radiocarbon methods thus providing a record of the timing of the onset of local erosion. Tallis's work on this topic is focussed on the heavily eroded peatlands of the Southern Pennine range in northern England.

Tallis (1997a and b) summarizes much of this work and suggests that in the Southern Pennines two main phases of gully erosion can be identified. The first, starting between 1250 and 1450 AD which is coincident with, or immediately postdates the Early Medieval Warm Period, is particularly associated with development of dendritic gully networks from existing hummock and pool topography. A second period of enhanced peat erosion is recognized post circa 1750 when there was considerable headward extension of gully systems into the peat mass. These patterns, together with the observation that within a gully system the dates of onset of erosion tend to get younger upstream (Tallis 1997b and c), emphasize the fact that gully erosion is an ongoing process rather than a particular event.

A second approach to the dating of erosion phases has been to investigate the depositional record in lake sediments downstream from eroded peatland landscapes. Rhodes and Stevenson (1997) studied seven lakes across Ireland and western Scotland with evidence of former peat erosion in their catchments. Rapid increases in the organic content of lake sediments, assumed to represent the onset of catchment peat erosion, date to between 900 and 1800 AD, with the majority erosion episodes occurring in the period 1500–1800 AD. The study rejects fire as a general cause of the erosion because there is no statistical relation between charcoal records from the lake cores and peat erosion. The relatively early onset of erosion also suggests that intensification of grazing and atmospheric pollution were unlikely to be the primary triggers. Rhodes and Stevenson suggest that the concentration of erosion episodes during the Little Ice Age (1500–1850) implies that more severe climatic conditions during this period are an important control on the onset of erosion.

In Ireland Bradshaw and McGee (1988) studied lake sediment sequences in Wicklow and in Donegal. Increases in organic content measured using loss on ignition together with evidence of reversal of radiocarbon dates was used to determine the onset of catchment erosion. The data suggest that catchment erosion began 3,000 years BP (radiocarbon years before present) in Wicklow but that erosion did not begin until 1500 BP further west in Donegal. The early dates of erosion recorded at two widely spaced sites suggest that natural processes rather than any anthropogenic impact were the key controls on the initiation of erosion.

It is clear that significant progress has been made towards the identification of major periods of onset of peat erosion. In the UK the evidence points to peat erosion being a phenomenon largely of the last millennium although earlier dates are recorded in Ireland (Figure 1.7). Dating erosion is however only part of the process of arriving at an apparent cause for the initiation of erosion. The main approach to identifying cause from the palaeoenvironmental record has been through the correlation of the onset of erosion with other known periods of environmental change whether from the historical record or reconstructed from proxy evidence. In this respect Tallis's work on intact mire sequences has significant advantages over the lake sediment evidence in that it tells us in considerable detail how peat surfaces develop and provides evidence of the mire surface mechanisms involved. However this approach cannot categorically identify cause. The onset or acceleration of peat erosion identified from several regions over the last 250 years (Mackay and Tallis 1996; Rhodes and Stevenson 1997; Tallis 1997b; Huang 2002) is coincident with intensification of upland agriculture, particularly sheep grazing (Shimwell 1974; Huang 2002), harsher climatic conditions in the Little Ice Age and impacts of atmospheric pollution on upland vegetation (Ferguson et al. 1978), all of which have been identified as potential causes of peat erosion.

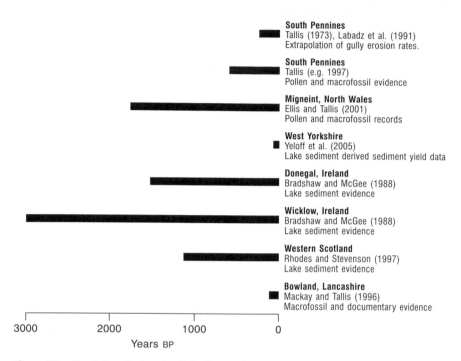

Figure 1.7   The timing of inferred periods of onset of peat erosion in the UK and Ireland

1.4.2   Direct observation of the onset of erosion

An example of the very rapid onset of erosion through instantaneous transformation of the bog surface vegetation is by wildfire events. Anderson (1997) reports a catastrophic fire on Burbage Moor in the Southern Pennines which damaged vegetation across 120 hectares. This fire occurred in the very dry summer of 1976 and in the subsequent September unusually high rainfall led to stripping of up to 1 metre of peat exposing the mineral surface beneath. Similarly Maltby et al. (1990) describe erosion of peat on the North York Moors of northern England in the aftermath of fires in the summer of 1976. Here significant removal of peat directly through combustion, through water erosion but also by deflation produced a surface of exposed mineral substrate and gullies in exposed peat. There is clear evidence that fire can produce dramatic and rapid erosion of upland peats (Radley 1960), and with the total areas of bare peat far exceeding locations where only gully erosion is dominant. In locations where natural vegetation regeneration is impaired, through for example pollution or grazing pressure, fire scars can be persistent features in the landscape (Anderson et al. 1997). In areas with relatively high burning frequency the aggregation of numbers of persistent fire scars in the landscape can make them a significant proportion of total erosion (Anderson et al. 1997). However because the impact of fire is local it is not a sufficient explanation for the widespread onset of erosion unless there is evidence of climate change likely to significantly increase fire frequency in moorland sites.

1.5   Causes of Peat Erosion

The initiation of peat erosion is a complex process which may be triggered by a variety of different impacts. Evidence shows that the dates of initiation of peat erosion in the UK are spread across the last millennium (Figure 1.7). Therefore, rather than search for specific causes of peat erosion in particular times and places, an alternative is to consider the onset of peat erosion as a threshold process (Schumm 1979). At the threshold the mire system switches from an intact system state to an erosional state, where rates of material flux from the system (including water and solute flux as well as sediment) and the principal controls on those fluxes are significantly altered. The mire system is in many ways analogous to badland systems (Tallis 1997a) where the friable peat layer is protected by a dense 'caprock' of vegetation. The initiation of erosion is controlled by the balance of the forces of erosion (frost, wind, rainfall, runoff) and the ability of the vegetation layer to resist erosion. Shifts between the two system states can therefore be produced either by increases in the erosive force or by a reduction in strength of the vegetation layer. The former is the mechanism invoked by Stevenson et al. (1992) in suggesting that colder, wetter and stormier Little Ice Age climates were central in triggering erosion. The latter encompasses the wide range of impacts of the mire surface which tend to stress the vegetation layer including fire, overgrazing, pollution, desiccation and trampling. Over time external changes, such as change in climate, and land management, shift the balance between the eroding and resisting forces potentially triggering erosion by crossing an extrinsic (externally forced) threshold.

It has also been argued that some local areas of intense erosion are a result of crossing intrinsic thresholds. Tallis (1985a) suggests that some erosion of blanket peat in the English Peak District has been triggered by marginal mass movements which he attributes to peat instability due to natural peat accumulation beyond a critical depth (Chapter 5). Nevertheless it is clear that a significant proportion of peat erosion is a response to external forcing of the mire system, removal of vegetation and exposure of bare peat surfaces to the elements. Overall it is possible to conceptualize

Table 1.4 Trigger factors for peat erosion on Holme Moss (after Tallis 1997b)

Date	Impact	Effect
1450	Desiccation Medieval Warm Period	Initiation of gully erosion
1770	Major fire	Produced extensive bare peat areas
1770	Marginal peat slides	Exposed bare soil and rock downslope
1800	Loss of *Sphagnum*	Reduced peat formation
1940	Overgrazing	Exacerbation of erosion
1976	Fire	Further bare peat areas
1983	Television mast construction	Major disturbance of mire surface

eroding peat landscapes as lying on a spectrum of erosion potential controlled by local climate and land use. Many of the external impacts are highly spatially variable, leading to a complicated mosaic of intact and eroding peat surfaces. This is particularly well illustrated by Tallis (1987; 1997b), who uses the example of heavily degraded blanket peat at Holme Moss in the Southern Pennines to illustrate the multiple triggers for erosion at a single site (Table 1.4).

The fact that we cannot readily identify regionally consistent causes for peat erosion should not be seen as detracting from the importance of the work that has been done to identify contributory factors. The question of causation has direct implications for the management of eroding peatland landscapes. Much of the legislatively defined value of blanket bogs is tied up with the unique vegetation and faunal populations which they support. This has led to re-vegetation of eroding peatlands as a conservation strategy with the focus as much on re-vegetation as an end in itself as on re-vegetation as an erosion control strategy. This restoration focus is appropriate if it is demonstrated that the onset of erosion is due to human intervention in a natural system. If in fact at least some part of the peat erosion observed in the landscape has natural origins then it can be argued that the gullies and remnant peat islands of an eroding peatland are a distinctive mire surface microform (Lindsay 1995), and merit conservation as part of the spectrum of natural mire surface conditions (see Chapters 8 and 9).

1.6   A Brief History of the Evolution of Peatland Geomorphology

Despite a relatively small body of literature, investigations of the geomorphology of peatlands have a long history. This brief chronology is not an attempt at an exhaustive review but rather an illustration of the development of the field. It relates to observation and measurement of the forms and processes of erosion, rather than the environmental conditions which predispose peatlands to degradation. The overall aim is to provide context for the work reported here.

1.6.1 Accounts of erosion in the natural science tradition

The earliest documentary references to peat erosion typically describe the dramatic erosional effects of rapid mass movements in peatlands (see Chapter 5).

These accounts are very numerous and range from scientific descriptions such as this account of a peat slide at Port Stanley on the Falkland Islands (Mulvaney 1879: 803):

> During the night of the 30th November 1878 there occurred a phenomenon of a most unusual type in the Falkland Islands, – an avalanche of peat which nearly overwhelmed the chief settlement. The peat bogs on the heights above Stanley, the chief town, gave way and the black oozy mud rolled down the hill with a momentum that neither the iron stanchions around the reservoir nor the barriers by the sea could withstand . . .

through to poetic (but surprisingly detailed) descriptions such as this account of the Crow Hill (near Bradford, UK) bog burst of September 24th 1824:

> But the summers heat the heaps of peat
> Had dry'd in many a gaping chink
> And when so dry the clouds on high
> Send down a flood to give it drink
> And as each flaw with greedy jaw
> Quaft with unsatiated thirst
> The lightenings flashed, the thunders crasht
> And its tremendous bowels burst
> (Verses three and four of 'The Phenomenon', a poem on the Crow
> Hill bog burst by John Nicholson, Ogden, 1976)

1.6.2 Descriptive accounts of widespread peat erosion

The first references to extensive erosion occur in the early ecological accounts of British moorlands. An excellent review of these early descriptive accounts is given by Bower (1962) spanning early work on Scottish peat bogs by Aiton (1811), Geikie (1866), Lewis (1905) and Crampton (1911), later work on the Peak District blanket peats by Moss (1913), through to the influential first modern accounts by Pearsall (1950,

1956). These studies were based on description of the extent and form of erosion and largely treated this as an interesting variation on the ecology of the mires. The most detailed classification of eroding moorlands was undertaken by Bower (1960a; 1960b; 1961; 1962) who mapped erosion across the Pennine moorlands of northern England. Bower's work was highly influential but in common with much of the geomorphological work of the first half of the twentieth century inferred the nature of peat erosion processes from observation of the resultant landforms. As a consequence, qualitative interpretations of the dominant erosion processes varied between authors. Bower's work emphasized fluvial processes whereas Radley (1962) studying South Pennine moorlands suggested that aeolian processes were the principal cause of surface recession in areas of bare peat. Despite some attempts to reconcile these views (Barnes 1963) the quantitative measurements required to assess relative impacts of various processes were not available.

1.6.3   Quantitative observations of blanket peatlands

In common with many of the early descriptions of peatland geomorphology many of the early process measurements were made by ecologists. Crisp (1966) produced quantitative estimates of sediment yield from a small eroding peat catchment in the North Pennines and Tallis (1973) produced some of the first quantitative measurements of gully erosion in peat. These early studies began to provide estimates of the rates of gully development in eroding peatlands and calculated typical sediment yields, but there was relatively little attention on the processes controlling sediment flux. It was not until the 1980s that geomorphological studies of the processes of peat erosion were undertaken. Two important studies in particular began to examine the important role of sediment production on bare peat faces, as interpreted from temporal patterns of sediment export at timescales ranging from annual to the individual storm (Francis 1987; Labadz 1988; Francis 1990; Labadz et al. 1991). During the same time period extensive survey of reservoir sediments from catchments in the Southern Pennines began, giving a detailed picture of typical sediment yields from an eroding peatland region (White et al. 1996).

A second strand of quantitative work relating to the geomorphology of peatlands during the 1980s focussed on the material properties of peat and in particular the causes of peat slope failure. The widely cited work by Hobbs (1986) is still one of the best summaries of peat material properties. Carling (1986a and b) produced the first detailed process-based explanations of peat mass movements using observations of a series of slides in the North Pennines.

Over the past ten years a range of different approaches has been applied to further develop quantitative understanding of peat erosion processes. Yeloff et al. (2005) is one of the first studies to directly compare peat catchment sediment yields derived from reservoir sediments with the proxy records of catchment conditions (e.g. pollen) preserved in the lake sediment microfossil record. This approach has allowed more precise correlation of sediment flux with changing catchment conditions. A number of studies have also examined erosion processes at the sub-catchment scale on experimental plots, developing much needed understanding of the processes of sediment production, transport and deposition (e.g. Holden and Burt 2002c; Warburton 2003). At the same time there has been increased interest in upscaling and generalizing this understanding. Approaches have included the use of sediment budgets to examine in more detail connectivity in catchment sediment supply (Evans and Warburton 2005) and also the application of a range of remote sensing technologies (e.g. Haycock et al. 2004; McMorrow et al. 2004). Of these perhaps the most significant development is the availability of high resolution DEMs (2-metre scale) derived from LiDAR (Light Detection and Ranging) laser altimetry which offer the potential to explore the nature and topographic associations of gully erosion and provide the prospect of developing models of peat erosion applicable at the landscape scale (see Chapter 4, Figure 4.5).

These changing paradigms in peatland geomorphology reflect wider changes in the focus of geomorphology over the last century, from an early interest in descriptive landscape studies, through a period of detailed and quantitative process studies at the small scale, and culminating in contemporary interest in applying the quantitative understanding gleaned from this work to explain geomorphological change at landscape scales. Church (2005) provocatively suggests that modern geomorphology has become divided into two camps, one engaging with the 'scientific' problems of large-scale earth system science and the upscaling and generalization of present understanding of earth surface processes, and the other preoccupied with the application of this understanding at a local level to solve problems of environmental management. Both these tendencies can be recognized in geomorphological work on peatlands. An interest in environmental management has been a significant thread in peatland geomorphology and the need to preserve physical integrity of upland surfaces as a prerequisite for conservation of their ecological diversity has underlain much research. However, the potential importance of peatlands as carbon stores means that an understanding of peatland geomorphology is important in contemporary debates over climate change and the role of peatland carbon budgets (Worrall et al. 2003). There is a large literature on the role of dissolved organic carbon production (Couper et al. 1997) and export from peatlands (e.g. Waddington and Roulet 1997; Freeman et al. 2001a and b), but surprisingly, given the scale of the sediment flux from eroding sites, relatively little work on the role of erosion and consequent particulate carbon flux in the carbon cycle.

One of the aims of this volume is to summarize current knowledge of the physical processes, and to identify areas for future research required both as an input to global change debates but also to provide a scientific underpinning to much of the experimental practical conservation being undertaken on eroded moorland. Because of the importance of peatland preservation and the potentially rapid rates of change, management of the erosion of upland peats involves landscape-scale intervention, so that effective management of the environment and understanding of the controls on and trajectories of the evolution of the land surface are closely intertwined. Therefore for the geomorphology of peatlands at least the dichotomy of focus suggested by Church (2005) needs to be resisted. It is undesirable, if even possible, to separate the process knowledge required as a contribution to understanding global carbon balances from the conduct of local management of eroded moorland surfaces.

1.7   Structure of this Volume and the Peat Landsystem Model

Conceptual models of peatland landscapes tend to emphasize the hydro-ecologically determined forms of intact mires (e.g. Figures 1.1 and 1.2). In eroding peatlands the surface morphology is heavily influenced by erosional and depositional landforms at a range of scales. Figure 1.8 is a representation of the peat landsystem. It illustrates a series of common geomorphological forms of upland peatlands. Surface forms common to intact mires such as pools and hummocks are shown, as well as a series of erosional/depositional features. These include features associated with mass failure of the peat, the characteristic forms of gully erosion, collapsed pipe systems, eroded drainage channels and areas of peat deposition at the interface between peat slopes and upland river systems. Not all of these features are necessarily a feature of all upland systems, similarly the features presented may be present in a range of degrees of development. For example gully systems may range from shallow incipient erosion within the peat to severely eroded but re-vegetated systems which may have the morphological characteristics of an eroded system but a well developed moorland vegetation cover. Nevertheless the geomorphological functioning of most upland peatlands could be summarized using a selection of these features and an assessment of their degree of development. This conceptual model of the landforms and sediment system linkages in peatland environments is a necessary first step in the process of constructing an empirical sediment budget (Dietrich et al. 1982). The remainder

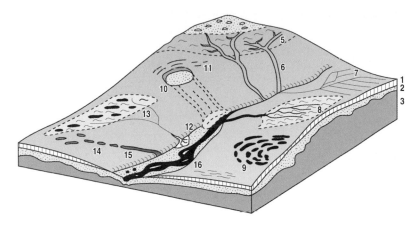

1	Peat deposits	9	Bog pool complex
2	Glacial/periglacial deposits (substrate)	10	Peat mass movement
3	Bedrock	11	Peat tears and tension cracks
4	Deflation surface remnant peat hummocks	12	Valley side peaty debris fan
5	Gully (Type I)	13	Eroded pool and hummock complex
6	Gully (Type II)	14	Collapsed pipe system
7	Artificial channels (grip network)	15	Peat block sedimentation
8	Peat haggs	16	Upland river system (mineral sediment)

Figure 1.8   The upland peat landsystem. Schematic representation of the range of features to be found in upland peatland landscapes. Not all features will be necessarily present at a particular site

of this volume aims to elucidate the details of this landsystem. Central to understanding the physical functioning of mire systems is the flow of water through and across them, hence Chapter 2 reviews the current state of knowledge of the hydrology of upland mires. Chapters 3–6 address key geomorphic processes operating in peatlands (sediment production, fluvial erosion, slope processes and wind erosion) and Chapter 7 identifies the morphological expression of the combination of erosion processes operating in upland mires. Chapter 8 examines the interaction of erosional and ecological processes and the consequences of erosion at the landscape scale. Finally, Chapter 9 explores some of the implications of widespread peat erosion, and provides conclusions.

Understanding peatlands is a fundamentally interdisciplinary endeavour (Charman 2002). Geomorphological understanding has advanced steadily over the past half century, but has been secondary to the large volumes of ecological and hydrological work. Our objective is that the summary of geomorphological understanding presented in the following chapters will give the geomorphological perspective on peatland functioning wider prominence.

Chapter Two

The Hydrology of Upland Peatlands

2.1   Introduction

Blanket peats are typically 90 per cent water. It is therefore perhaps unsurprising that there is intimate linkage between the hydrology and geomorphology of these systems. The hydrology of peatlands has been the subject of several comprehensive reviews (Ingram 1983; Egglesmann et al. 1993). The aim of this chapter is not to replicate these but rather to focus on the hydro-geomorphology of peatlands and the particular aspects of their hydrology which are closely linked to the form of blanket peat landscapes, and processes of sediment transfer. The chapter is divided into two main sections; the first considers the basic controls on the hydrology of blanket peatlands with a particular emphasis on recent research, and the second considers the geomorphological implications of the characteristic hydrology of the systems.

2.2   Controls on Water Movement in Peat Landsystems

2.2.1   Hydraulic conductivity of upland peat soils

Peat as an organic sediment has unusual geotechnical properties including compressibility, low density and its fibrous nature. Many of the challenges of peatland hydrology and geomorphology relate to these unusual properties and the fact that standard measurement techniques often require modification to yield acceptable results for peat soils. Thorough reviews of peat physical properties are provided by Hobbs (1985) and Egglesmann et al. (1993). From a hydrological perspective the key physical property of peatlands is hydraulic conductivity. In some senses peat-

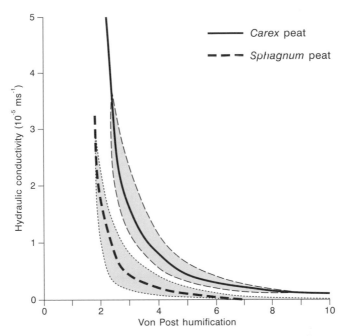

Figure 2.1   Relationships between peat degree of humification and hydraulic conductivity in *Sphagnum* and *Carex* peat (after Egglesmann et al. 1993)

land hydrological systems are relatively simple and many aspects of their hydrology follow logically from understanding patterns of variability of hydraulic conductivity within the peat mass. Typically the physical properties of peat soils are highly variable. The source of this variability is both spatial and temporal variation in the plant communities which provide the parent material for peat formation, and also in the soil-forming processes (largely humification and mineralization of plant material) which produce the stable accumulations of organic sediment which comprise upland peatlands. In a typical peatland there are characteristic changes in the character of the peat with depth. These changes are best understood in the context of the decomposition of peat over time and consequently the increasing degree of peat humification with depth. As plant tissue is decomposed, largely through the action of aerobic bacteria, the ratio of humic substances to plant tissue increases and the average pore size within the peat decreases (Boelter 1965; Egglesmann et al. 1993). Correlated with decreases in porosity are decreases in the bulk density, water content and the hydraulic conductivity of the peat. Figure 2.1 shows the relationship between the degree of peat decomposition and hydraulic conductivity after Egglesmann et al. (1993). These data suggest an apparently simple relationship between increased decomposition, decreased pore size in the peat and consequently reduced hydraulic conductivity. Peat soils are compressible so that changes in porosity are not simply associated with peat depth but will also occur with change in loading. Hemond and Goldman (1985) have shown that measured hydraulic conductivities in peat can vary with the applied head of water, and hence the loading on the peat, and argue that the cause of this effect is compression. Holden and Burt (2003a) applied compressible soil theory to the measurement of hydraulic conductivity through monitoring head recovery in dipwells in upland blanket bog. They showed that the resulting hydraulic conductivity (k) values were five times less than those obtained assuming a rigid soil. These results emphasize the potential importance of the compressible nature of peat soils. Price (2003) has demonstrated that the hydraulic conductivity of peats decreases as the water-table drops and suggests that the cause is compression of the peat collapsing large pores. Compression occurs because the effective stress on the peat column is increased as water-table falls due to decreases in the buoyancy provided by pore water pressure. The effect is most marked in the surface layers where moisture content is reduced, but is observable through the whole profile including below the water-table (Figure.2.2).

2.2.2   The diplotelmic mire hypothesis

The vertical changes in peat properties resulting from progressive decomposition do not occur gradually with depth, rather there is a sharp distinction between a relatively low-density upper layer and a much higher density lower layer. In the Russian literature a distinction has long been made between an active upper layer and an inert lower layer (e.g. Ivanov 1981). Ingram (1978) formalized this classification in English literature identifying the upper layer as the acrotelm and the lower as the catotelm.

The characteristics of the two layers as identified by Ingram are summarized in Table 2.1. The definition hinges on the position of the water-table. In the upper acrotelm layer plant litter is supplied from the bog vegetation. Fluctuations in water-table allow periodic access of air to the pores so that litter is rapidly decomposed through the action of aerobic bacteria and fungi. The abundance of fresh litter and relatively undecomposed nature of the upper layers mean that the density of the acrotelm peat is relatively low and the hydraulic conductivity relatively high. However, because of the rapid decomposition in this zone the hydraulic conductivity declines and density increases rapidly with increasing depth.

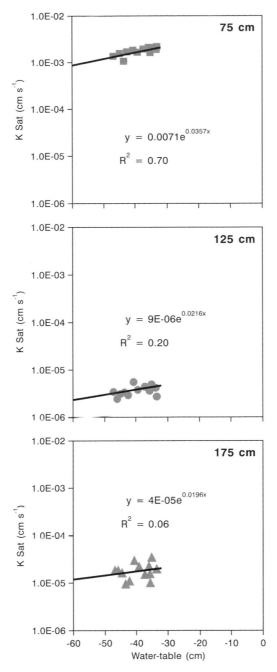

Figure 2.2 Changes in hydraulic conductivity (k) with water-table depth at three peat depths in an undisturbed plateau bog near Lac St Jean (after Price 2003)

Table 2.1   Characteristics of the acrotelm and catotelm (Ingram 1978)

	Acrotelm	*Catotelm*
Water-table	Fluctuating	Permanently saturated
Aerobic status	Periodically aerobic	Anaerobic
Moisture	Variable, intensive exchange of moisture with atmosphere and surrounding area	constant
Hydraulic conductivity	High, declining rapidly with depth	Low
Water yield	High	Low
Exchange of energy and matter	Fast	Slow
Microbial activity	High numbers and activity, aerobic and anaerobic	Low numbers and activity, anaerobic

Beneath the long-term average minimum water-table depth the permanent saturation of the catotelm produces anaerobic conditions which suppress microbial activity and slow decomposition. The dense peats of the catotelm have a hydraulic conductivity which can be 3–5 orders of magnitude lower than the acrotelm.

This two-layer 'diplotelmic' classification of the peat profile has been widely adopted as a useful description of the functioning of many bog systems. Implicit in many discussions of acrotelm peat is the role of *Sphagnum* in the hydrological functioning of this layer. Holden and Burt (2003b) note that Ingram and Bragg (1984) describe the acrotelm as a circa 50-centimetres-deep layer of *Sphagnum* peat. Holden and Burt however use field data to demonstrate that if peat horizons are characterized hydrologically rather than ecologically, the diplotelmic mire hypothesis is also a useful description of hydrological processes in mires where *Sphagnum* is rare or absent. The basic model is therefore useful in degraded or eroded mires as well as undisturbed systems, but Holden and Burt (2003b) caution against simple two-dimensional applications of the concept. Given the increased topographic complexity of eroded mire systems this is an important caveat.

Figure 2.3 shows data from a raised bog and a blanket bog in North America which demonstrate the typical pattern of vertical-change hydraulic conductivity (k). Values of k in the acrotelm are several orders of magnitude above those in the catotelm decreasing with depth. Hydraulic conductivity within the catotelm is much lower and bears no systematic

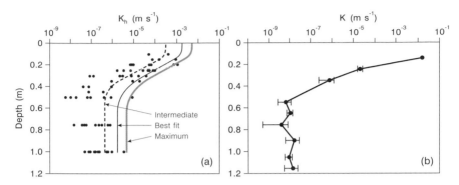

Figure 2.3   (a) Variation of hydraulic conductivity with depth in the Mer de Bleu bog, Ontario (after Fraser et al. 2001). (b) Vertical variation in k in blanket bog at Cape Race, Newfoundland (after Hoag and Price 1995)

relation to depth. Although the general pattern of decreasing k between the acrotelm and catotelm is common to most studies vertical changes in k within the catotelm appear to be more varied. Beckwith et al. (2003a) show a log linear decrease in k with depth in the catotelm for samples from a cutover raised bog in Yorkshire, UK whereas Clymo (2004) shows a steady decrease to depths of 5 metres followed by a general increase in samples from Ellergower Moss in southwest Scotland. Such variation is perhaps unsurprising. In the acrotelm the parent material of the peat is reasonably constant and increasing decomposition and compaction with depth are the key controls on peat density and hydraulic conductivity. Catotelm peats however may represent several thousand years of peat accumulation. Therefore although a general increase in density and decrease in k with depth might be expected due to compaction at depth, variations in parent material might override this effect. Changes in bog surface vegetation associated with long-term climate change mean that peat type is likely to vary through the catotelm. Similarly long-term changes in moisture produce variation in the rates of peat growth and hence degree of decomposition of catotelm peat (Belyea and Clymo 1998).

Hydraulic conductivity is a key parameter in modelling groundwater flow within peatlands. Most reported field hydraulic conductivity data are derived from head recovery methods based on the time required for water-table measured in dipwells to recover its former level after being pumped out (e.g. Boelter 1965; Hoag and Price 1995; Kennedy and Price 2005). Table 2.2 demonstrates very considerable heterogeneity in the hydraulic conductivity of upland peats in both space and time. This pattern is further complicated by the fact that hydraulic conductivity commonly

Table 2.2   Typical values of hydraulic conductivity reported for ombrotrophic mires

Location	Peat type	Hydraulic Conductivity cm s^{-1}	Reference
Mer de Bleue	catotelm	10^{-3}–10^{-5}	(Fraser et al. 2001)
Mer de Bleue	acrotelm	10^{-1}–10^{-3}	
Thorne Moors Raised bog	acrotelm	10^{-2}–10^{-3}	(Beckwith et al. 2003b)
	catotelm	10^{-3}–10^{-5}	
Lac St. Jean, Quebec	catotelm	10^{-2}–10^{-5}	(Kennedy and Price 2005)
Cape Race, Newfoundland blanket bog	acrotelm	1	(Hoag and Price 1995)
	catotelm	10^{-5}–10^{-6}	
Moor House blanket bog	catotelm	10^{-5}–10^{-7}	(Holden and Burt 2003a)
Review of various blanket peats	catotelm	10^{-5}–10^{-8}	(Rycroft et al. 1975)
Ellergower moss	catotelm	10^{-6}	(Clymo 2004)

demonstrates anisotropy, i.e. differences in the horizontal (h) and vertical (v) components of k. (Boelter 1965; Schlotzhauer and Price 1999; Beckwith et al. 2003a) with k_h typically exceeding k_v. Mean anisotropy ratios (log10 k_h/k_v) observed by Beckwith et al. (2003a) for raised bog samples were 0.55 although significant vertical variation was observed.

2.2.3   Groundwater flow in upland peatlands

Groundwater flow in peats is commonly considered as a diffusive process (Reeve et al. 2001). As such it is controlled by a gradient and by a constant which in the case of groundwater flow is formulated in Darcy's Law.

$$Q = -k\frac{d\mathrm{h}}{d\mathrm{l}}$$

where Q is the discharge per unit area (specific discharge), k is the hydraulic conductivity and $d\mathrm{h}/d\mathrm{l}$ is the hydraulic gradient or head difference h over distance l.

Darcy's Law is used to describe flow through a matrix of water-filled pores between peat particles. Ingram (1982, 1983) argues that because of the saturated nature of lower peats this is an appropriate model of groundwater flow in peat. Others (Rycroft et al. 1975a and b; Hemond and Goldman 1985; Waine et al. 1985) have argued that, particularly at lower hydraulic conductivities, water flux through peat will be non-Darcian in nature, although Baird et al. (1997) argue that non-Darcian behaviour is only significant at high hydraulic gradients which are unlikely to be found in natural systems.

Darcy's Law implies that for a given peatland, average water flux is determined by patterns of k. As previously discussed (Section 2.2.2) vertical variation in k is important. Baird et al. (1997) argue that in some peatlands the differentiation of acrotelm and catotelm by hydraulic conductivity is less pronounced than the diplotelmic model assumes. They note that as the catotelm is typically an order of magnitude greater in depth than the acrotelm, differences in k of an order of magnitude would result in an equal volume of discharge from surface and deep peats. Were this the case there would be a much greater degree of water storage and release within upland peat systems than is commonly observed (see Section 2.2.5). Nevertheless the association between the climatic history of the peatland and the nature of the deposited peat means that there may be distinct stratigraphic layers with enhanced hydraulic conductivity. In particular the fibrous and less humified recurrence surfaces associated with periods of bog surface inundation provide the potential for significant water flux in deeper peats.

Darcy's Law offers an intuitively attractive and apparently simple approach to modelling groundwater flows within peatland systems. In practice however both the three dimensional pattern of k and the anisotropy of k discussed above significantly complicate the simple application of diffusion models to peatland groundwater flow (Baird et al. 1997). Beckwith et al. (2003a and b) have drawn attention to the importance of considering both the anisotropy and heterogeneity of hydraulic conductivity within a peat mass and demonstrated that heterogeneity in particular can significantly alter the nature of modelled flowpaths.

It is commonly assumed that the dominant vector of groundwater flow is parallel to the ground surface as, due to the horizontal stratification of peat deposits, lateral hydraulic conductivity typically exceeds vertical. Some studies have however demonstrated that under particular conditions there may be an important vertical component of groundwater flux. Glaser et al. (1997) demonstrate that in raised bogs in Minnesota although the long-term average water flux is down, during drought periods there is upward movement of groundwater recharging the depleted upper layers. Reeve et al. (2000, 2001) used a finite difference approach to modelling

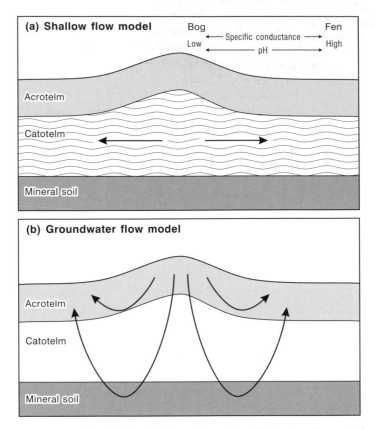

Figure 2.4   A conceptual model of peatland groundwater flow. Shallow lateral catotelm flow dominates where the underlying mineral stratum is relatively impermeable whilst higher permeability allows deeper vertical water flux (after Reeve et al. 2000)

groundwater flow and suggest that the key control on vertical water flux is the permeability of the peat, with lateral flow dominating in locations where the underlying mineral soil is impermeable, and a greater importance of vertical water flux where mineral soil permeability increases (Figure 2.4).

Clymo (2004) reports velocities of lateral water movement in a Scottish mire above an impermeable basal layer as low as $0.6 \, \text{mm a}^{-1}$ due to very low hydraulic conductivities. This very low rate implies that the influence of deep peats on the quantity and quality of peatland runoff is very small. Clymo suggests that significant variability in hydraulic conductivity both within and between bogs is typical so the relative importance of matrix groundwater flow will also vary. Lapen et al. (2004) presented modelling results which suggest that in blanket bog in Newfoundland lateral drainage is impeded particularly by more humifed peats of lower hydraulic conductivity around the mire margins. Recent work has suggested that one important control on the hydraulic conductivity of peats and consequently on rates of groundwater flow is the production of gas (CH_4) bubbles within the catotelm. Gas bubbles have the potential to block pores and reduce the hydraulic conductivity of peats (Reynolds et al. 1992; Strack et al. 2005). Beckwith and Baird (2001) have demonstrated a 5–8-fold reduction in hydraulic conductivity of blanket peats from northern England after laboratory incubation and gas accumulation. The action of the bubbles has been implicated in observations of non-Darcian flow through peats (Ingram et al. 1974). Increasing recognition of the importance of gas bubbles combined with the difficulties of accurate measurement of k and the highly variable patterns of k in bog peats mean that the intuitively attractive approach of modelling peat groundwater flows based on Darcy's Law appears to be rather simplistic.

Groundwater dynamics are fundamental to understanding the nature of ombrotrophic mires but perhaps counter-intuitively the key characteristic of groundwater flow in many ombrotrophic mires is that the water flux is very low. This is the characteristic which allows maintenance of water-tables above the level of those in surrounding systems. Even though quantitatively groundwater exchanges may be small there is important further work to be done on understanding peatland groundwater dynamics.

Recent recognition of the degree of anisotropy and heterogeneity in hydraulic conductivity in peats, the role of gas bubbles and interactions with macropore flow raise significant challenges of measurement and for the modelling of peatland groundwaters. A proper understanding of the groundwater system is vital because in raised mires there is an intimate connection between groundwater dynamics and the overall form of the mire (Ingram 1987; see Section 2.3), and also because groundwater flow is a distinctive flowpath within the peatland hydrological system with implications for outflow water quality and the consequent role of peatlands in global biogeochemical cycling. Given that the discharge from catotelm peats is very low, the exchange of moisture between the mire and surrounding hydrological systems is dominated by evaporation and near-surface runoff which are discussed further in the following sections.

2.2.4 Evaporation

One consequence of the very low rates of groundwater flow within upland peatlands is the relative importance of water flux to the atmosphere by evapotranspiration as a mode of water loss from the mire system. During periods of low rainfall when low baseflows illustrate the limited extent of groundwater flow, moisture loss from the peat mass is dominated by eva-potranspiration. This is clearly illustrated by the widespread observation (Heikurainen 1963; Boatman and Tomlinson 1973; Ingram 1983; Evans et al. 1999) of a diurnal pattern to bog water-table recession. Steeper water-table recession during daylight represents a response to increased rates of evapotranspiration and stabilization of water-table at night reflects limited transpiration during the hours of darkness (e.g. Figure 2.5). Typi-cally the scale of these diurnal 'steps' is on the order of 10–20 millimetres, although in Tasmanian blanket mires daily fluctuations of up to 300 millimetres have been reported (Pemberton 2001).

Evaporation from mires has been reviewed thoroughly by Ingram (1983) and Egglesmann et al. (1993). Table 2.3 presents reported rates of evaporation from ombrotrophic mires around the world which post-date these reviews. Noteworthy is the remarkable consistency of reported evap-oration rates from a wide range of mire types and geographical locations. This may reflect both the limited range of climatic conditions within which ombrotrophic mires develop (Lindsay et al. 1988) and the common condition of these mires that water-table is high for most of the year. The apparent consistency of average rates however masks local variability in important controls on evaporation. For example in Newfoundland Price (1991) notes that frequent fog reduces evaporation with rates dropping from $2.5\,\mathrm{mm\,d^{-1}}$ to an average of $1.1\,\mathrm{mm\,d^{-1}}$ on foggy days. In New Zealand Campbell and Williamson (1997) note strong stomatal control exerted by the restiad vegetation so that some of the lowest evaporation rates occur in mid-summer. The extent and nature of stomatal control on evapotran-spiration across a range of bog types is an important research area since projected climate warming will tend to drive higher evaporation rates from ombrotrophic mires and this is likely to be an important factor in lowering water-tables and consequent mire degradation.

2.2.5   Runoff generation

'When all is said and done peat has a great virtue not often realised: that of storing water in a catchment area and releasing it continuously long after rain and floods have ceased' (Pearsall 1950: 370). Historically this has been a widely held view of the hydrological functioning of peatlands which is frequently repeated in the literature. There is however very little evidence to support this view. Some early work, such as the comparative catchment studies of Conway and Millar (1960), demonstrated that intact *Sphagnum*-dominated catchments showed longer times to peak, and Hall

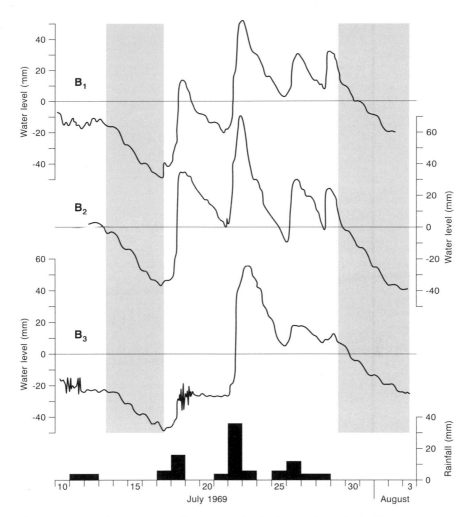

Figure 2.5   Diurnal steps in water-table recession at Brishie Bog, southwest Scotland. Note in particular the highlighted areas of the graph (redrawn after Boatman and Tomlinson 1973)

and Cratchley (2005) also report a buffering effect of *Sphagnum* cover. However, although there may be a measurable effect of *Sphagnum* cover in delaying flow in some catchments this is only a relative change. All of the catchments monitored by Conway and Millar (1960) showed a very flashy runoff. Bragg (2002) have also demonstrated delays in flood peaks from a Scottish raised mire on the order of 24 hours, but they ascribe this to the buffering effect of the lagg fen rather than the processes of runoff

Table 2.3   Reported rates of evaporation from ombrotrophic mires

Location	Evaporation rate $(mm\,d^{-1})$	Reference
European Russia	1.5–2.8	Kurbatovdiv 2002
Central Siberia	1.5–3.3	Kurbatovdiv 2002
Northern New Zealand raised bog	1.4–3.8	Campbell and Williamson 1997
Newfoundland blanket bog	1.1–2.5	Price 1991
New Zealand raised bog	dry day average 2.7	Thompson et al. 1999
UK (North Pennine) blanket bog	average 1.5	Evans et al. 1999
Quebec	1.9–3.6	Van Seters and Price 2001
Central Sweden	1–3.5	Kellner and Halldin 2002

production in the ombrotrophic mire. In fact the evidence from the preceding discussion of very low hydraulic conductivities in bog peats together with the high water-tables characteristic of the system mean that there can be very little storage within most upland mires. It is now widely accepted that the sponge analogy is misleading (Ingram 1987; Baird et al. 1997; Bragg 2002; Holden and Burt 2003b).

Runoff from peatlands is typically flashy with storm hydrographs exhibiting short lag times, fast times to peak and rapid recession to low base flow levels (Figure 2.6). The typical annual flow regime of peatland streams is dominated by individual storm events and consequently closely mirrors the rainfall pattern for the year (Figure 2.7). Rapid rates of hydrograph recession and the year-round incidence of very low flows between storm events are consistent with the very low hydraulic conductivity of peats and the consequent limited contribution of peat groundwater discharge to low flow maintenance. It should also be noted that much of the work on peatland runoff relates to eroded or drained mires which may be more productive of surface runoff (Baird et al. 1997), although Holden et al. (2004) argue that the evidence on this point is equivocal. Whilst drainage and erosion may enhance an already flashy runoff regime, the flashy nature of runoff from upland peatlands is a general characteristic of upland peatlands and is related to the dominant processes of runoff generation. Burt and Gardiner (1984) and Burt et al. (1990) suggested that measured infiltration capacities for blanket peatlands in the Southern Pennines of as low as $2\,mm\,hr^{-1}$ indicated that infiltration excess overland flow may be generated on moorland sites. The infiltration capacity was apparently related to the mosaic of moorland vegetation with *Eriophorum* cover exhibiting particularly low values (Figure 2.8). Contrary results

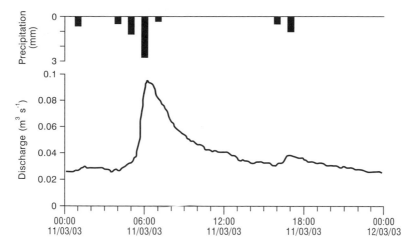

Figure 2.6   Typical storm hydrograph for a peatland stream from a 0.4 hectare blanket peat stream on the southern flanks of Bleaklow, South Pennines. Note rapid time to peak and complete rise and recession within 24 hours

were presented by Evans et al. (1999) who demonstrated that in the North Pennines the water-table was at or above the surface 83 per cent of the time and that significant stream runoff occurred only in these periods. The North Pennine study catchment included areas of both erosion and drainage but the rapid runoff from the system was generated through saturation overland flow as rain fell onto already water-logged surfaces. Rainfall occurring when the water-table was depressed did not produce runoff and stream flow was delayed by groundwater recharge. However the system exhibited very low storage. Figure 2.9 illustrates water-table recovery and runoff generation after a severe drought in the summer of 1995 and illustrates that even after this extreme event flow was only delayed by 12 hours. The typical response of the system to rainfall was rapid generation of saturation overland flow. Similarly Hall and Cratchley (2005) report that water storage in upland blanket peats in northwest Wales is very limited, with saturated conditions developing within a single storm event. The importance of saturation overland flow has been confirmed for the North Pennines by tension infiltrometer measurements (Holden et al. 2001) and by rainfall simulation (Holden and Burt 2002c). It seems likely that although there may be local generation of infiltration excess overland flow in intense convective storms, the runoff response of most ombrotrophic peatlands is controlled by saturation overland flow during the frequent low-intensity frontal storms. This is particularly the situation in the case of the amphi-Atlantic blanket peatlands where the

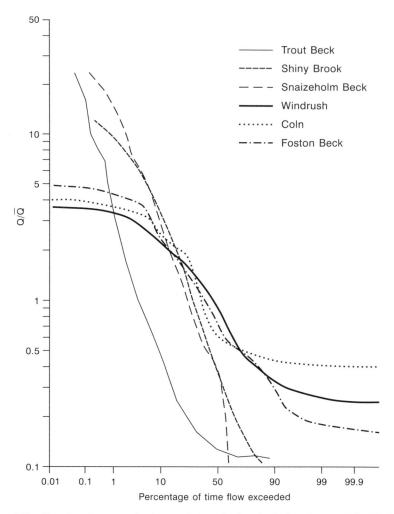

Figure 2.7   Flow duration curves for three typical small mineral soiled catchments (Coln, Windrush, Foston Beck) and three catchments with predominantly peat soils (Trout Beck, Snaizeholm Beck and Shiny Brook) (redrawn after Burt [1992] with additional data from Evans et al. [1999])

climate is predominantly oceanic and lower-intensity frontal rainfall is dominant.

Although runoff is routinely measured in hydrological studies of peat-lands there has been surprisingly little work on the mechanisms of runoff generation. Much of the available evidence comes from eroded or drained mires and might therefore be unrepresentative of wider northern wet-lands. However, as the effect of drainage would be to draw down water-

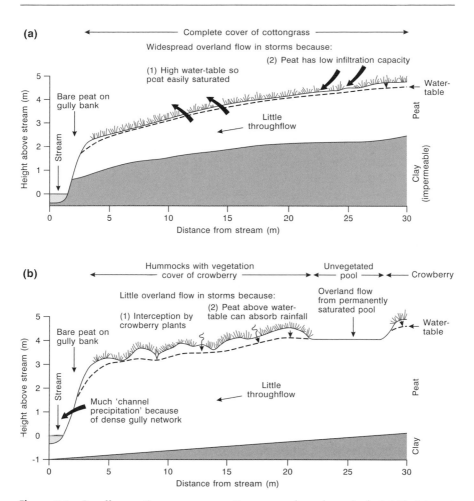

Figure 2.8 Runoff generation processes on cottongrass- and crowberry-dominated blanket peat moorland (redrawn after Burt and Gardiner [1984])

tables, observations of rapid generation of saturation overland flow in these systems suggest that this mechanism should be dominant in undamaged mire systems. The common character of upland bogs is that they saturate rapidly and are highly productive of runoff.

Aggregate catchment behaviour is consistent with rapid wetting up of upper layers of the peat and the generation of saturation overland flow, but the high hydraulic conductivities of upper peat layers mean that a significant contribution to runoff is also made by the development of shallow subsurface stormflow through the acrotelm (Burt et al. 1990, 1997).

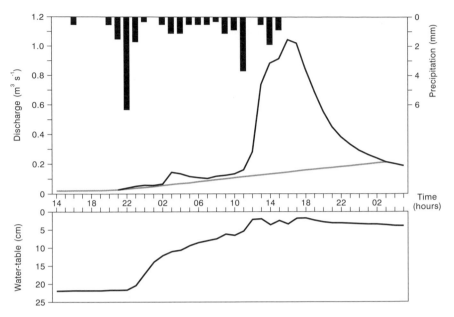

Figure 2.9   Runoff generation on Trout Beck in the North Pennines after a severe drought in 1995. Initial rainfall raises the water-table and significant streamflow is delayed until the water-table is near the surface allowing generation of saturation overland flow (after Evans et al. 1999)

Soulsby et al. (2003) emphasize the importance of subsurface stormflow in contributing to runoff in blanket peat catchments in northeast Scotland but do not distinguish relative proportions of overland and near-surface stormflow. Holden and Burt (2003b and c) demonstrate that in fact these proportions are variable in space and time using data from blanket peats in the North Pennines (Figure 2.10). The partitioning of runoff between various flow pathways in upland peats has been investigated in more detail in a series of studies of runoff generation in blanket peatlands in the North Pennines (Holden and Burt 2002a and b, 2003a and c). These demonstrate that overland flow and near-surface stormflow are the dominant mechanisms in these systems. They account for 96 per cent and 81 per cent of runoff generation under high flow and low flow conditions respectively. Saturation overland flow is dominant in generating runoff on the rising limb of storm hydrographs whilst drainage of the acrotelm is demonstrated to be important in generating runoff during the recession limb of the hydrograph. Partitioning of subsurface stormflow and overland flow is under topographic control with near-surface flow important on steeper slopes, and overland flow (generated either as return flow or saturation

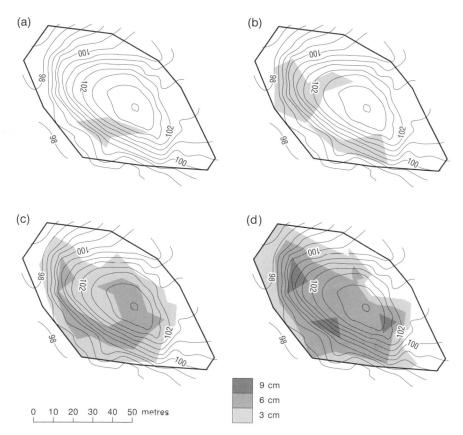

Figure 2.10   Evolution of depth of overland flow at 6-hour intervals (a–d) during a storm on a North Pennine blanket peat slope (after Holden and Burt 2003b)

overland flow) dominant in saturated footslope locations (Holden and Burt 2003b). The basic pattern of runoff generation is therefore consistent with the partial area model of Hewlett and Hibbert (1967).

Natural soil pipes are a ubiquitous feature of UK blanket peatlands (Holden 2005). These macropores, which range in scale from centimetres to metres, provide rapid subsurface drainage routes and comprise an important part of the hillslope hydrological system in these uplands. Holden and Burt (2003b and c) found matrix through-flow from the lower layers of the peat to be an insignificant component of stormflow, but note that on average up to 10 per cent of runoff can be generated from pipes within the deeper peat. In some storms pipeflow may contribute up to 30 per cent of recession flow. These pipes show a rapid stormflow

Table 2.4 Water-table studies on ombrotrophic mires

Location	Period	Precipitation (mm)	Runoff (mm)	Evaporation (mm)	Change in storage (mm)	Reference
Massachussets, USA, natural sphagnum bog	1976–1977	1,448 (53%)	246 (17%)	1,019 (73%)	+183	Hemond 1980; Charman 2002
North Pennines, UK, eroded and re-vegetated blanket peat	Sept 1994–Sept 1996	1,925	1,362 (72%)	534 (28%)	+27	Evans et al. 1999
Central Sweden, natural sphagnum bog	May–October	255	58 (23%)	271 (77%)	–75	Kellner and Halldin 2002
St Arsene Bog, Quebec, natural sphagnum bog	May–August 1998	212	0	281 (100%)	–58	Van Seters and Price 2001
Cancouna bog, Quebec, re-vegetated harvested bog	May–August 1998	220	54	285 (129%)	–100	Van Seters and Price 2001
Blanket bog, Newfoundland, Canada	16 May–24 June 1990	254	160 (63%)	71 (28%)	–8	Price 1992
Mer de Bleue bog, Ontario, Canada	May 1998–May 1999	757	222 (29%)	598 (79%)	–63	Fraser et al. 2001

response indicating that the macropore pathways link to the surface and flow in these pipes bypasses the soil matrix. In shallow peats in mid-Wales almost 50 per cent of runoff is bypassing flow through the pipe network (Jones and Crane 1984; Jones 2004). Similarly Carey and Woo (2000, 2002) have demonstrated that piping contributes up to 21 per cent of runoff from permafrost affected peat soils in the Yukon, Canada.

In summary, upland peatlands are systems characterized by high water-tables and a low water-storage capacity. Rapid runoff generation through saturation overland flow and near-surface stormflow through the highly permeable acrotelm dominate the runoff response. Flow from deeper peats is largely bypassing flow from well-developed pipe networks which is an important component of runoff, particularly on the falling limb of the hydrograph. Very low hydraulic conductivities in catotelm peats mean that runoff production from throughflow in deep peats is minimal. Consequently storm hydrographs are characterized by rapid recession and very low flows develop rapidly during drought periods.

2.2.6 The water balance of ombrotrophic mires

Upland peatlands are wetland systems and maintenance of high water-tables is central to their formation and functioning. Ultimately therefore water balance is the fundamental physical control on peatlands.

The water balance of ombrotrophic mires can be represented as follows

$$Q = R - E + \Delta S$$

Q = runoff, R = rainfall, E = evaporation, ΔS = change in storage

Detailed measurements of the water balance of mires are relatively rare. Table 2.4 lists a series of studies which have attempted to estimate the main parameters over various monitoring periods. The table shows significant variability in the relative proportions of water loss accounted for by runoff and evapotranspiration. Some, but not all, of this variability is accounted for by differences in measurement periods since most mires tend to have a period of water deficit in the summer and recharge in the winter (e.g. Hemond 1980; Evans et al. 1999). It is notable that the bogs in the table where runoff is the dominant loss are blanket bog systems and this may reflect the requirement for high precipitation inputs required to initiate bog growth on the sloping uplands typical of blanket peatlands.

The equation for the water budget presented above assumes that there are no exchanges of groundwater within the mire, and this is an assumption often made in studies of ombrotrophic mires. For many upland

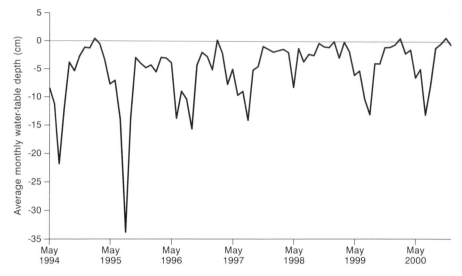

Figure 2.11   Annual patterns of water-table on blanket bog at the Moor House Nature Reserve, North Pennines (modified after Worrall et al. 2004)

systems these are small components but evidence of vertical movements of groundwater within the Mer de Bleue bog (Fraser et al. 2001) suggest that this is not always a reasonable assumption. The data demands of a complete water budget are very high and calculation of a budget often requires compromise either in the measurement period, the techniques of measurement or in initial assumptions. It is therefore important that potential errors in water budgets are clearly identified. Kellner and Halldin (2002) present a formal error analysis of a water balance for an intact Swedish mire and suggest that errors associated with the budget are on the order of 5 per cent for precipitation, 20 per cent for runoff and 20–30 per cent for evaporation. The relatively large error for evaporation reflects the difficulty of estimation of this parameter. Table 2.4 contains estimates based on Penman calculations (Evans et al. 1999), Bowen ratio measurements (Kellner and Halldin 2002) and lysimeter measurements (Van Seters and Price 2001).

The key parameter in the long-term health of mire systems is the change in storage. On an annual basis there is considerable change due to summer drawdown (Figure 2.11). In stable conditions the summer deficits are replenished by winter recharge so that the long-term annual mean change in storage is zero. Persistent negative changes in storage indicate an overall drying of the bog and a downward trend in water-table and may be associated with drainage, periods of drought or climate change. Change in storage is often calculated as the residual of the water balance and so is particularly influenced by error in measuring or modelling the main components. Measured in this way, it is therefore not a reliable indicator of long-term change in bog hydrological status. An alternative approach is to consider the water-table within the mire as an indicator of storage volumes so that long-term changes in water-table reflect changes in storage (e.g. Figure 2.11). This approach is widely adopted as a proxy for the net water balance of the mire but it should be regarded only as an approximation. Kellner and Halldin (2002) illustrate that for an intact mire in central Sweden changes in water-table only accounted for 60 per cent of change in storage with the rest accounted for by expansion and contraction of the peat mass with the addition of moisture, so-called mire breathing (Ingram 1983; Schlotzhauer and Price 1999).

The challenge of completing and closing the water budget even for ombrotrophic mires (which are probably the simplest mire type in water budget terms) are very significant. However, this is an important area of research since water balance is fundamental to understanding mire functioning, and a proper understanding of mire dynamics requires not only quantification of the water-balance components but exploration of the nature of their interactions.

2.3   Geomorphology and the Hydrology of Upland Peatlands

Within a typical upland peatland system the morphologically-related macrotopes and mesotopes (cf. Lindsay [1995] hydro-topographic units) are controlled by the pre-existing topography of the landscape. Blanket peat rarely exceeds 4 metres in depth except where it covers pre-existing basin peats so that the peat blanket smoothes rather than obscures underlying topography at large scales. Graniero and Price (1999) have demonstrated that topographic parameters could explain 22 per cent of the variance of bog and heath patterning in Newfoundland blanket bogs but that the key parameter is the topography of the pre peat surface rather than the present day peatland topography. They also argued that downslope areas were favoured for blanket bog formation due to delivery of runoff from upslope bog surfaces. Similarly Turunen and Turunen (2003) have demonstrated rapid spread of peat accumulation in sloping bogs in British Columbia on rolling gentle topography. Graniero and Price (1999) argue that their data support the spread of peatlands both by upslope spread of peat formation and by downslope spread from upland plateaux. The implication is that the development of the peatland system is topographically and hydrologically controlled with initial accumulation centres located in areas of lower relief and consequently reduced drainage.

A key functional component of mire systems which is commonly over-looked is the perennial drainage system which often divides distinct hydro-topographic units within the mire. Major drainage lines can be inherited from the pre-peat topography as the development of peat cover will tend to increase the flashiness of the hydrological system increasing peak flows in the major outlet channels. Consequently increase in stream power and potential lateral erosion of the peat mass (Evans and Warburton 2001) will tend to preserve the course of the pre-peat drainage line. In headwater locations the balance between the spread of peat-forming vegetation and stream-erosive power may be reversed so that there may be complete overgrowth of pre-existing drainage lines as described by Thorp and Glanville (2003).

One part of blanket peat complexes is isolated from direct control of the pre-existing topography. Raised mires, within upland mire complexes, are often located at initially topographically-determined accumulation centres, but vertical peat growth creates an isolated hydrological system. Since the topography of these features is entirely determined by the form of the peat mass the form might be expected to be related directly to hydrological processes. Ingram (1982) proposed a geophysical explanation for the topographic form of raised mires widely known as the 'groundwater mound hypothesis'. In this model ecological processes are secondary to physical ones in determining the form of raised mires. Essentially the model argues that impeded drainage due to low hydraulic conductivity in the catotelm produces a hemi-elliptical domed water-table the height of which can be determined analytically from the mean hydraulic conductivity of the mire and the rate of lateral discharge. Ingram (1982, 1987) has demonstrated that the model provides good prediction of the form of two mires in Scotland. The model originally proposed by Ingram makes a number of simplifying assumptions, most importantly that the hydraulic conductivity of the mire is constant. Several studies have demonstrated that significant spatial (Kneale 1987) and vertical (Clymo 2004) variability in k, together with pronounced anisotropy of k (Beckwith et al. 2003a), mean that the assumption of homogeneous values of k which is required to produce an analytical solution for mire form does not hold. Various authors have produced alternative models which incorporate spatially and vertically variable values of hydraulic conductivity (Armstrong 1995; Bromley and Robinson 1995) and such numerical-simulation approaches appear to offer the best hope of accommodating more realistic peat properties and more complex mire forms (Clymo 2004). The refinement of Ingram's approach has not resulted in any challenge to the fundamental premise that the large-scale morphology of raised mires is physically controlled by the low rates of water transmission through catotelm peats.

Upland peatlands are wetland systems with high water-tables as their defining physical characteristic. It is therefore an unsurprising conclusion that their morphology is closely linked to their hydrological function. However, the nature of that control varies between mire forms. The preceding section argued that, at the level of the macrotope and mesotope, mire morphology is hydrologically controlled directly in raised mires or indirectly through topographic location of peat accumulation centres in other components of upland mire complexes.

At the level of the microtope, identification of key controls on mire topography is more controversial but ecological processes increasingly appear to be important determinants of microtopography. At this scale the key control on mire surface topography is the role of differential rates of productivity and decomposition on formation of hummocks and hollows and on pool formation and development (Hulme 1986; Svensson 1988; Belyea and Clymo 1998; Glaser 1998; Belyea and Clymo 2001). An excellent summary of recent work in this area is given by Charman (2002). Charman argues that in non-permafrost-dominated peatlands surface patterning is dominated by biotic processes. The interaction of hydrological and ecological processes on the mire surface is complex and forms the basis for much of the current literature of peatland science. Some recent work for example argues that pool patterning on mire surfaces can be explained simply through spatially variable patterns of near-surface hydraulic conductivity in hummocks and hollows (Swanson and Grigal 1988; Couwenberg 2005). Most work, however, explicitly or implicitly takes mean water-table to be the key hydrological parameter for explanation of mire surface topography (Belyea and Clymo 2001). Water-table influences productivity and decomposition through its control of both litter decomposition rates and influence on dominant mire surface species. The shift from mesotope to microtope therefore effectively marks a transition from dominance of mire form by hydrology and pre-peat topography to dominance by hydrological and ecological factors. Indeed Lindsay (1995) distinguishes the two scales in blanket peat complexes as hydro-topographic units and hydro-ecological microforms.

In the particular case of eroding mires a further set of geomorphological microforms comprises the major small-scale relief on the bog surface (see Chapter 7). The nature of these geomorphological forms and the processes controlling their development and distribution is the subject of Chapters 3–7. Central to much of this discussion is the importance of water movement through and over upland peats as a driver of erosion.

Groundwater and macropore flow within deep peats is important in generating locally high-pore water near the peat-mineral interface and promoting landsliding of the peat mass (Carling 1986a; Warburton et al. 2004; Chapter 5). It has also been suggested that in some cases the onset of gully erosion in upland peats is associated with the expansion and eventual unroofing of large pipes within the peat mass (Bower 1960b; Holden et al. 2002; Jones 2004). The rapid generation of saturation over-land flow on upland mire surfaces (Evans et al. 1999; Holden and Burt 2003b) also plays an important role in the origin and development of gully erosion in degrading peatlands both through initial incision of mire sur-faces and removal of weathered peats from bare surfaces (see Chapters 3 and 4).

The focus of this section has been on the role of hydrological processes in determining the geomorphology, particularly the morphological form of mire systems. Of course landform morphology is an important deter-minant of hydrological function, so that, particularly in eroding mires, there are significant feedbacks between geomorphological change and hydrological change. Development of extensive gully networks on mire surfaces has the effect of increasing drainage density and consequently increasing the hydrological connectivity of hillslopes and channels. The catchment studies of Conway and Millar (1960) demonstrated that peat-land systems in northern England showed increased water yield and flashiness of runoff in eroded and drained catchments, and Holden et al. (2006 in review) suggest that this effect is particularly marked in areas of natural gullying. Development of gully networks within a catchment also creates natural drainage of the areas close to the gully lines leading to lower and more variable water-tables (e.g. Figure 2.12). Lowered water-tables impact on the quality of water draining from the peatland, and in particular the deepening of the acrotelm and consequent oxidation of surface peats is likely to produce increased levels of dissolved organic carbon and associated pollutants (Clark et al. 2005; Rothwell et al. 2006). The consequences of peat drainage also have a potential effect on the physical hydrology of the system. Desiccation of surface peats can lead to development of hydrophobicity and an increased importance of macropo-res as flow pathways (Holden and Burt 2002b). Holden et al. (2006 in review) demonstrate a clear association between incision of peatlands and the development of extensive piping. The development of significant mire surface topography through erosion therefore has the potential to signifi-cantly modify both the locus of runoff generation (through changes in the location of saturated partial areas) and the pathways via which runoff is delivered to the main channel.

Upland peat landscapes are wetland systems. The importance of under-standing their hydrological functioning has long been recognized as central to understanding and managing their ecological functions (Ingram 1983; Bragg 2002). The remainder of this volume focuses in more detail on the physical processes of erosion of upland peat and on the interaction

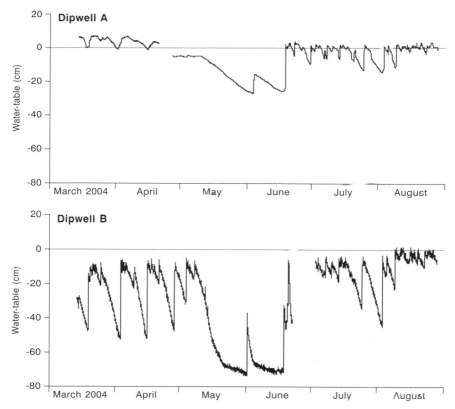

Figure 2.12   Water-table data from two dipwells in Southern Pennine blanket bog. Both dipwells are in the Upper North Grain catchment, a 40 hectare catchment on the southern slopes of Bleaklow. (a) Dipwell A is in an area of relatively intact peat. (b) Dipwell B is immediately adjacent to a 2-metre-deep gully. Deeper average water-table and greater variability occur at the gully edge

of these physical processes with ecological processes which is central to the restoration/regeneration of degraded upland peatlands. The role of flowing water on and through the peat plays an important role in this story.

Chapter Three

Sediment Production

3.1 Introduction

The production of sediment on hillslopes is an important link between hillslope hydrology and fluvial sediment flux. The aim of this chapter is to assess current understanding of the nature of sediment supply from bare peat surfaces to the channel system in peatland catchments. There are two main parts to the discussion. The first reviews measured rates of slope erosion (sediment supply). These are important in understanding gully erosion, longer-term landform development, and constraints on sediment flux to the fluvial system. The appropriate interpretation of these data in terms of sediment production is also investigated. The second presents some new data and synthesizes the limited work on mechanisms of sediment production to develop an understanding of the process controls on sediment supply.

3.1.1 Monitoring sediment production using erosion pins

One of the most widely used techniques for the measurement of surface erosion is erosion pins (Haigh 1977) and the technique has commonly been applied to erosion on bare peat surfaces (e.g. Imeson 1974; Tallis and Yalden 1983; Anderson et al. 1997; Evans and Warburton 2005). Weathered material is rapidly exported from peat slopes because of the low density of the peat. The system is therefore typically supply-limited so that the surface retreat rates measured by erosion pins are inferred to be a reasonable approximation of the rate of sediment supply to the fluvial system. Erosion pins (typically wooden dowels or thin metal rods, <5 mm diameter and <600 mm length) are inserted into an eroding surface and surface retreat is monitored through measurement of the length of exposed pin. A number of factors can affect the accuracy and precision of erosion pin measurements. Couper et al. (2002) identify four main potential sources of error which are:

1 movement of the pins;
2 changes in surface elevation;
3 influence of the pin on erosion;
4 human interference.

All of these problems are potentially relevant to erosion pin measurements of the retreat of peat surfaces. Movement of the pins is most likely through frost heave during the winter, or potentially through swelling of the peat as it re-wets. If the pins are long and well anchored it is possible that the pin remains still whilst surface peat layers expand and contract leading to apparent changes in surface elevation. The most common approach to minimizing error from this source has been to re-measure pins seasonally under similar surface conditions (e.g. Anderson et al. 1997). In peat the interaction of the pins with the peat is largely controlled by the peat type. In fibrous peat pins can interfere with transport of peat across sloping surfaces and 'overhangs' of peat develop above the pins. This is less of a problem in well humified peats and can be minimized through the use of thinner pins (2 mm diameter or less). Given the remote and inhospitable nature of many peatland landscapes human interference is not a widely reported problem, but interference from grazing animals, particularly sheep has been reported from the UK where sheep grazing is common in areas of peat moorland (Yang 2005). In any erosion pin study the precision of the technique, and hence the useful temporal resolution is limited by these sources of error. In practice this means that erosion pins are better suited to measurement of surface retreat at low temporal resolutions. Figure 3.1 illustrates that for a site where mean retreat rates are 19 mm a^{-1} precision of ±20 per cent is achieved within two years. After two years the error associated with erosion estimates levels off and it can be assumed that after this time reliable estimates of average rates can be determined. This is a useful guideline for considering the minimum length of time over which monitoring should take place.

Reported rates of surface peat erosion based on erosion pin measurements are presented in Table 3.1. These data all represent at least one year of measurement, and the reported rates are relatively consistent, falling with a single exception, in the range 5–45 mm a^{-1}. The reported rates are extremely high and illustrate the potential for rapid removal of peat cover once the vegetated surface is removed. Interestingly there is as much variation within

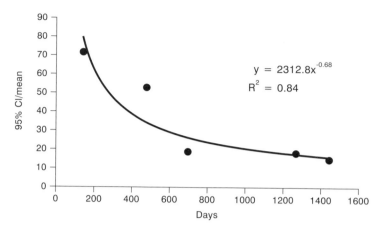

Figure 3.1   Change in the 95% confidence interval (CI) of mean erosion pin retreat as a proportion of the mean over time. Mean data points from 79 erosion pins installed in bare blanket peat on the Moor House National Nature Reserve in the North Pennines, UK, monitored over four years, 1998–2002

the South Pennine measurements, which represent the largest single geographical concentration of measurements, as there is within the rest of the dataset. There is also a tendency, as previously noted by Tallis and Yalden (1983), for lower angled surfaces to have higher rates of erosion. It is therefore likely that the extent of erosion is highly variable at small spatial scales controlled by local topographic context. Potential mechanisms controlling this variability are considered later in this chapter.

3.1.2   Sediment trap data

One of the effects of the considerable measurement error associated with the erosion pin approach is that it is difficult to obtain precise estimates of erosion at short timescales. An alternative approach to measurement of erosion on peat faces is the use of sediment traps. Figure 3.2a illustrates a Gerlach trough (Gerlach 1967) type device constructed from guttering and pinned to the peat face. The trap includes a shield to minimize losses from rainsplash and wind action. Sediment traps aggregate erosion from an area of the peat face rather than at a point, and the collected sediment can be accurately quantified gravimetrically, therefore sediment traps can produce reliable data at much higher resolution than the point measurements of erosion pins. This is potentially important; whereas annual measurement provides useful information about *rates* of retreat monthly, weekly or per storm, measurement of surface erosion together with other

Table 3.1 Reported rates of surface retreat measured on bare peat by erosion pins

Location	Context	Period (years)	Surface retreat rate (mm a^{-1})	Reference
Macquarie Island, Tasmania	Low-angled peat surface	3.3	43	Selkirk and Saffigna 1999
Moor House, North Pennines	Gully walls	4	19.3	Evans and Warburton 2005
Shetland Islands	Summit peat	5	10–40	Birnie 1993
Upper North Grain, South Pennines	Gully walls	3	14	Unpublished data
Plynlimon	Hagg faces	5	30	Robinson and Newson 1986
Snake Pass, South Pennines	Gully walls	1	7.8	Philips et al. 1981
Moor House, North Pennines	Gully walls	1	10.5	Philips et al. 1981
Holme Moss, South Pennines	Low-angled peat margin	2	33.5	Tallis and Yalden 1983
Holme Moss, South Pennines	Peat Margin	1	73.8	Philips et al. 1981
Harrop Moss, Pennines	Bare peat surface	7	13.2	Anderson et al. 1997
Snake Pass, South Pennines	Peat margin	1	5.4	Philips et al. 1981
Mid-Wales	Ditch walls	1.4	23.4	Francis and Taylor 1989
North York Moors	Low-angled bare peat surfaces	2	40.9	Imeson 1974
South Pennines	Low-angled flats	1	18.4–24.2	Anderson 1986
Cabin Clough, South Pennines	Low-angled eroded face	2	18.5	Tallis and Yalden 1983
Doctors Gate, South Pennines	Low-angled eroded face	2	9.6	Tallis and Yalden 1983
Plynlimon, Wales	Peat faces	2	16	Francis 1990
Forest of Bowland	Summit peat	1	20.4	Mackay 1993

Table 3.2 Cumulative sediment trap contents from eroding gully walls in the Rough Sike catchment, Moor House National Nature Reserve, North Pennines, UK (for site locations see Figure 3.2b) July 1999–July 2000

Site*	1	2	3	4	5a	5b	5c	5d	5e
Area M² ◇	0.3	0.35	0.4	0.65	1.3	1.4	0.3	0.25	0.225
Slope ◇	90	70	80	90	75	28	90	90	90
Aspect	S	E	S	N	E	E	E	E	E
Sediment catch (g)	407.3	137.3	142.0	550.7	132.7	394.7	220.5	329.2	300.6
Equivalent retreat rate (mm)	14	4	4	8	1	3	7	13	13

* Location in Figure 3.2b.

Figure 3.2   (a) Location of Rough Sike Sediment traps. One trap at sites 1–4 and five traps at site 5. (b) 50-cm Gerlach trap with Splash shield at site 2. The traps are affixed to the peat face with aluminium pins

environmental data can potentially allow inference of *mechanisms* of retreat. The latter approach is developed later in the chapter. Sediment traps have been less widely applied to peatland systems than erosion pins; it is therefore important to establish the extent to which measured rates are comparable. Table 3.2 presents sediment trap data from a series of sediment traps monitoring five gully-side locations (Figure 3.2b) at Rough Sike in the North Pennines, UK. The sediment catch is converted to an equivalent surface retreat rate by normalizing by the area of exposed gully wall above the trap and applying a measured dry mass per wet volume value of $0.11\,g\,cm^{-3}$. The monitored retreat rates are in the range 1–14 millimetres for the annual monitoring period, and so are in the lower range of reported surface retreat rates monitored using erosion pins. Traps 3 and 4 have low sediment catches and also much the largest contributing areas. These traps are positioned at the bottom end of lower angled gully walls, and it is likely that because of lower slope angle and larger contributing area there is a transport-limited control on the trap sediment flux and some of the sediment produced is stored on the slope. These traps are therefore a less reliable indicator of sediment production from the high gully walls. The average retreat rate excluding these sites is $9\,mm\,a^{-1}$. This can be compared directly with erosion pin measurements from the same location which give surface retreat rates of $19.3\,mm\,a^{-1}$. Clearly the sediment trap measurements are not comparable to previous work using erosion pins. Comparisons of Gerlach trough and erosion pin derived degradation rates in badland areas of Spain (Sirvent et al. 1997) have suggested that significantly higher erosion rates are recorded by the troughs. It is therefore unlikely that the differences observed here are simply a function of the measurement technique. In blanket peat environments it is arguable that traps and pins are in fact recording different aspects of erosion.

One of the early debates over the nature of peat erosion concerned the relative importance of wind and water erosion (Bower 1961; Radley 1962). Francis (1990) has also highlighted the importance of peat wastage (i.e. oxidation of surface peats) in the retreat of peat faces. Erosion pins therefore measure the combined effects of three distinct processes; wind erosion, water erosion and chemical oxidation of the peat surface. In contrast sediment traps measure only the transfer of material across the peat face. In the Rough Sike case-study the sediment trap measurements equate to approximately 47 per cent of the total surface retreat measured by erosion pins. This ratio is consistent with reports by Yang (2005) from the Southern Pennines. The sum of measured wind erosion rates at the Rough Sike site (Evans and Warburton 2005) and the sediment trap contents is equivalent to 70 per cent of the surface retreat suggesting that on the order of 30 per cent of apparent erosion is accounted for by peat wastage. Table 3.3 after Evans et al. (2006) presents other published data which support the possibility that wastage accounts for a significant proportion of surface retreat. Such estimates are only approximations and further work is required on this topic (particularly in the light of the implications of significant carbon flux to the atmosphere from eroding peatlands, see Chapter 9), but it is likely that the sediment trap data are a better measure of sediment transfer to the fluvial system than erosion pin measurements.

Table 3.3 Potential degree of oxidation of surface peats. Peat wastage calculated as a residual (after Evans et al. 2006)

Site	Water erosion	Wind erosion	Peat wastage	Wastage %
Rough Sike[1]	$732\,\mathrm{g\,m^{-2}\,a^{-1}}$	$627\,\mathrm{g\,m^{-2}\,a^{-1}}$	$571\,\mathrm{g\,m^{-2}\,a^{-1}}$	30%
Upper North Grain	$1300\,\mathrm{g\,m^{-2}\,a^{-1}}$	$527\,\mathrm{g\,m^{-2}\,a^{-1}}$	$1573\,\mathrm{g\,m^{-2}\,a^{-1}}$	46%
Plynlimon				56–81%

Notes:
[1] Wind and water erosion are measured values derived from Evans and Warburton (2005).
[2] Water erosion at Upper North Grain is measured values from sediment traps (Yang 2005). No direct wind erosion data exist for Upper North Grain, as an approximation the Rough Sike value modified by the ratio of annual wind run between Upper North Grain and Rough Sike (0.84) is presented here.
[3] Plynlimon data derived from Francis (1990). The data are derived from locations exposed to and sheltered from wind/rain beat thus separating wastage and physical erosion effects.

3.2 Sediment Production as a Control on Catchment Sediment Flux

Upland peat catchments are by their nature relatively small headwater systems. Typically the sediment flux from such systems is intimately linked to the nature of sediment supply within the catchment (Knighton 1998). It might be assumed, given the apparently soft and erodible nature of peat and the extensive erosion documented in many blanket peat catchments (Tallis et al. 1997; Cooper and Loftus 1998), that restrictions on sediment supply would not apply to peatland systems. However, observations of sediment flux from peatland catchments appear to suggest that even in such apparently erodible catchments sediment supply is an important consideration and that catchment sediment flux may be supply limited.

Figure 3.3 shows a typical pattern of water and sediment discharge from a peatland stream. Peak discharge occurs ahead of peak flow so that there is positive hysteresis in the suspended sediment concentration (SSC)–discharge (Q) relation. The nature of the sediment supply control on fluvial sediment flux is considered in more detail in Chapter 4, but the important point here is that the usual explanation for positive hysteresis in SSC and Q relationships in small catchments is sediment exhaustion (e.g. Carling 1983; Francis 1990; Labadz et al. 1991; Kronvang et al. 1997; Slattery et al. 2002). Labadz et al. (1991) suggest that the observed

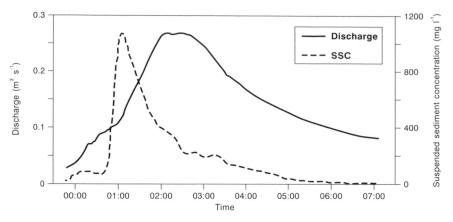

Figure 3.3   Positive hysteresis in the discharge-suspended sediment relationship from a small upland catchment in eroding blanket peat

sediment exhaustion is related to the importance of sediment 'preparation' or weathering as a control on sediment supply in eroding peatlands. The hypothesis is that freshly exposed peat is fibrous and resistant to water erosion, whereas weathering produces bare peat surfaces mantled with a layer of friable and easily erodible material. Two mechanisms have been widely identified in the literature as important in generating this surface layer; the action of frost and desiccation (Tallis 1973; Francis 1990; Labadz et al. 1991).

Frost is frequent in the cool upland climates that support blanket peatlands. Compared to mineral soils peat has a higher volumetric heat capacity but much lower conductivity and shows very marked contrasts in thermal response whether it is wet or dry (Fitzgibbon 1981). Significant temperature gradients can therefore develop at the peat surface. The

Figure 3.4   (a) A typical needle ice formation, note the sediment layer on the upper surface of the ice mass. Note also four layers of ice needles indicative of four consecutive nights of needle ice formation. Lawler (1993) suggests that depth of needle ice resulting from a single freezing event are in the range 2–31 millimetres, The hand (for scale) shows that each of these events was towards the upper end of this range. (b) Characteristic friable surface layer resulting from formation and thawing of needle ice. The surface material has a low density and a 'puffy appearance'. The frost detached material forms a distinctive layer above much denser undisturbed peat below. (c) A drying peat surface. Note the prominence of coarse fibres which bind the desiccating surface layer in a platy form

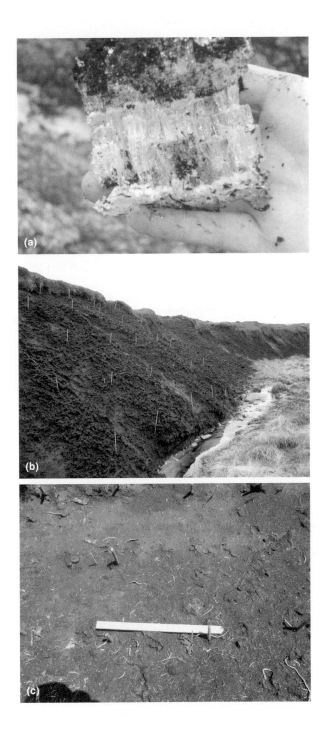

abundant supply of moisture and strong thermal gradients provide ideal conditions for the development of needle ice as water moves to the freezing front (Outcalt 1971). Due to the maritime location of many blanket peat environments, freezing is commonly diurnal rather than seasonal so the effect of a single needle ice event is multiplied many times through the winter season (Figure 3.4a and b). The importance of needle ice formation in eroding peat faces has been widely reported in environments ranging from alpine mires in Lesotho (Grab and Deschamps 2004) to erosion of peat remnants in Finnish Lapland (Luoto and Sepälä 2000), and is commonly observed in eroding upland peats in the UK (Tallis 1973; Legg et al. 1992).

The second mechanism commonly identified as important in the weathering of surface peats is desiccation (Burt and Gardiner 1984). In one sense this mechanism is closely related to the frost effects since the formation of segregation ice at the peat surface has the effect of desiccating the surface layer. The effects on the surface are however rather different. Desiccated surface layers develop over extended periods of dry weather and are characterized by platy aggregates of surface peat which are typically concave upwards and detached from the intact peat below (Figure 3.4c; Chapter 7). Polygonal desiccation cracking may also be visible on the surface (Francis 1990).

The 'weathered' material produced by each of these mechanisms is rather different. The fluffy peat produced by frost action is loose and granular and is transported to the stream as individual peat particles or fine aggregates of particles. Rainsplash and surface wash are important in transporting material across gully walls although Francis (1990) also reports dry gravitational transport with peat aggregates observed to 'detach and roll in strong gusts of wind' (Francis 1990: 450). The platy aggregates produced by desiccation weathering are very low density compared to the wet sediment made available by freeze thaw activity. The aggregates, as is common with dried peat, are hydrophobic (Ingram 1983). Once displaced from gully walls by wind or rain action they therefore tend to remain dry and large particles are transported relatively intact as flotation load in the stream.

There is considerable observational evidence that desiccation and frost action act to produce a superficial friable weathered layer on bare peat surfaces. It is plausible that the development of this layer is an important mechanism controlling sediment supply and produces the sediment exhaustion and characteristic positive hysteresis in peatland stream sediment rating curves. However there is very little direct evidence of the nature of sediment supply in these systems. In the following sections some field evidence of the role of climatic conditions in surface preparation is presented, and the limited evidence from rainfall simulation experiments

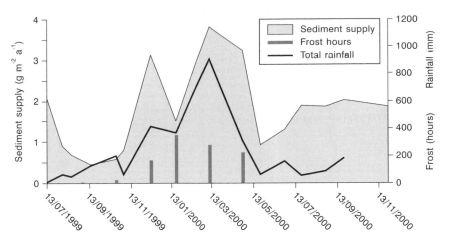

Figure 3.5   This illustrates sediment supply measured by traps over a 15-month period at Rough Sike. Highest rates of sediment production occur between January and March

is reviewed to identify the probable nature of the process link between sediment production and observed stream sediment flux.

3.3   Evidence from Field Observation

3.3.1   Climate correlations with trap data

Although it has frequently been suggested that the action of frost (Bower 1961; Tallis 1973; Tallis and Yalden 1983; Labadz et al. 1991; Evans and Burt 1998) and or desiccation (Tallis and Yalden 1983; Francis 1990) are key drivers of sediment production from bare peat surfaces there has been very little direct observation of the relation between climatic conditions and sediment supply. The Rough Sike sediment trap data referred to earlier in the chapter are collected at sufficient temporal resolution to reasonably allow comparison with climatic data. The sediment supply data are average sediment yields from the nine sediment traps. The traps were emptied approximately monthly over a 15-month period between July 1999 and November 2000. Actual collection intervals were constrained by access (average interval 32 days, minimum 11 days, maximum 69 days). Figure 3.5 and Table 3.2 illustrate that measured sediment trap contents vary considerably in time and the relatively large confidence intervals of the mean values indicate variability in space. However there is no clear pattern to the spatial variability; in particular there is no clear

Table 3.4 Correlation between climatic parameters and sediment supply measured as sediment trap catch for gully side sites at Rough Sike, North Pennines

	Minimum soil temperature 1 cm depth	Average Soil temperature 1 cm depth	Average soil temperature 10 cm depth	Total rainfall	Maximum rainfall intensity	Mean soil moisture	Frost hours	Frost hours per day
Pearson correlation	−0.510	−0.531*	−0.524*	0.718**	0.665**	0.540*	0.620*	0.540*
Significance	0.052	0.042	0.045	0.003	0.007	0.038	0.014	0.038

* Correlation is significant at the 0.05 level (2-tailed).
** Correlation is significant at the 0.01 level (2-tailed).

association with aspect which is the main source of between-site variation. Figure 3.5 plots the mean trap catch normalized for collection period as an indication of sediment supply from the bare peat walls of the gullies. A clear pattern is present of increased sediment supply in the winter months, in particular between November and April. This pattern would support the hypothesis that winter frost is an important control on sediment production. Continuous climate data is available from the study site via the Moor House Automatic Weather Station, which forms part of the Environmental Change Network (Sykes and Lane 1996). Figure 3.5 also plots total rainfall and total frost hours for the collection periods. This demonstrates an association between periods of high sediment supply and periods of frost and high rainfall. In order to test this association for each of the collection periods climate data relating to hypothesized controls on peat weathering (frost, soil moisture and rainfall) have been collated. Table 3.4 shows correlation data relating measured sediment supply to a range of derived climate indices. The strongest single correlation is a positive correlation with total rainfall which accounts for just over 50 per cent of the variance in sediment supply. There are also significant correlations with total frost hours and negative correlations with soil temperature. Forward stepwise regression of all the parameters against sediment supply identifies only total rainfall as a significant predictor. A combination of frost and rainfall parameters does not enhance prediction. It is therefore likely that the apparent correlation with frost activity is at least in part a function of the covariance between periods of frost and periods of higher rainfall apparent in Figure 3.5. It would be premature on the basis of this limited dataset to underestimate the role of frost in sediment production but it appears likely that the action of rainfall on the surface also has an important role to play in sediment production.

3.3.2   Direct observations of surface change

As has been discussed previously, studies of the rate of retreat of bare peat surfaces are relatively numerous but detailed process measurements relating surface conditions to the retreat of the bare peat faces have not been available. Figure 3.6 presents data from a monitoring experiment which aims to illustrate controls on the behaviour of the peat surface during erosion events.

The trace in Figure 3.6a is the daytime output from a photo-resistive erosion pin (PREP) inserted into an eroding peat gully in the Rough Sike catchment (Site 1 Figure 3.2). The PREP is essentially similar to the photo-electric erosion pins (PEEP) which have been widely used in monitoring erosion and deposition in fluvial contexts (Lawler 1991; Lawler

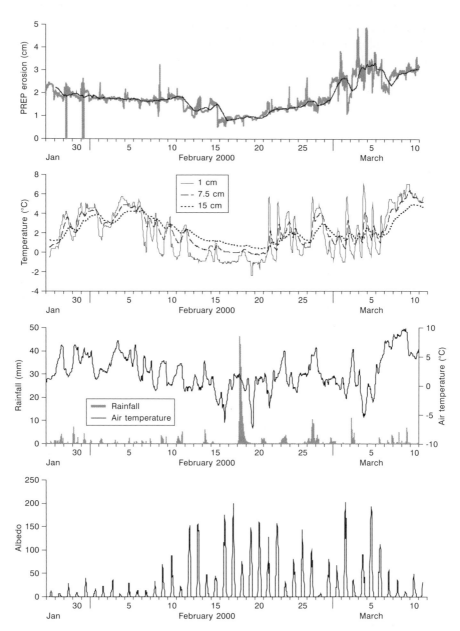

Figure 3.6 Surface retreat measured by photo-resistive erosion pins (PREP) (upper panel), together with local meteorological data for an eroding gully wall in the North Pennines, UK

et al. 2001; Lawler 2005). The pins are calibrated to provide a quasi-continuous readout of the length of exposed erosion pin which is recorded on a data logger. Plotted alongside the erosion pin data are meteorological data recorded for the catchment including rainfall, air temperature and surface temperature at three depths (1, 7.5 and 15 cm).

The plot covers the period late January to early March 2000. A period of frosty weather begins on 10th February and the albedo record suggests that this coincides with snowfall (albedo above 50% is interpreted as snowfall following Oke 1987). A steep temperature gradient develops in the upper layers of the peat conducive to needle ice formation. The depth of frost penetration is 7.5 centimetres which corresponds well with values of 4.5–7.2 centimetres reported by Francis (1990) in Welsh peats. For the first part of the data series until 12th February the length of exposed erosion pin is relatively constant at just under 2 centimetres. Between 12th and 21st February there is apparent surface advance and the length of exposed pin drops as low as 1 centimetre. This is interpreted as surface frost heave, the alternative possibility that the change represents snow accumulation is rejected as the south facing gully wall is vertical and sheltered, and the periods of surface advance do not correlate with high surface albedo values. Maximum surface advance of 1 centimetre occurs on February 17th, and from February 21st until the end of the series there is continuing surface retreat to eventual erosion pin exposure of 3 centimetres. The major snowmelt is recorded by the spike in rainfall recorded at the rain gauge on February 18th. The surface retreat is coincident with the onset of ground temperatures above freezing and continues over a period of two weeks during which there are several small rainstorms. This is interpreted as being indicative of removal of frost-weathered material produced in the preceding cold period by rainsplash and overland flow. The overall surface retreat of 10 millimetres is very high in the context of average annual rates of 19 millimetres at the site and suggests that severe winter frosts play a major role in preparation of erodible material and the erosion of peat faces.

3.4   Evidence from Controlled Experiments

Although the observation that intact fresh peat is relatively resistant to erosion has become conventional wisdom, the first experimental confirmation of this was presented by Carling et al. (1997). Through a series of flume experiments it was demonstrated that clear water velocities of $5.7\,m\,s^{-1}$ were required to produce erosion of a range of peat types, although erosion occurred at much lower velocities if there was mineral material suspended in the flow. Threshold velocities of $5.7\,m\,s^{-1}$ are clearly beyond

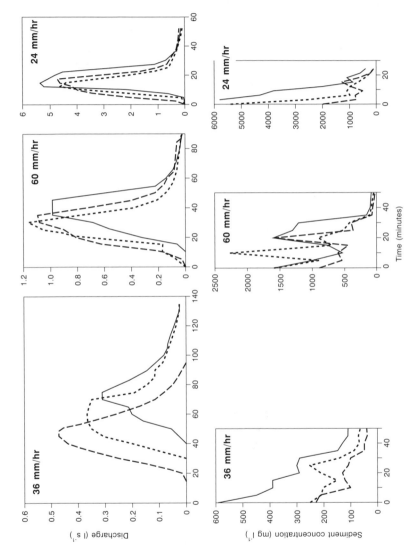

Figure 3.7 Sediment and water discharge from rainfall simulator plots under varying precipitation intensities and surface treatments. All plots show the same broad pattern of positive hysteresis in the runoff sediment concentration relation (after Klove 1998)

those likely to be produced in runoff from relatively low gradient peatland surfaces. These data appear to confirm that intact peats are highly resistant to erosion. They therefore provide strong supporting evidence for the role that weathering plays in the erosion of bare peat surfaces, particularly in the early stages of erosion on deep peat where there are very limited sources of mineral sediment.

There is very little existing experimental evidence on the mechanisms of sediment transfer to channels. Some of the more detailed data comes from a study of sediment transfer in peat mines in Finland (Klove 1998). In a series of rainfall simulation experiments Klove demonstrated that the rate of sediment removal from the mire surface increased non linearly as rainfall intensity increased. The measured runoff from the plots demonstrated positive hysteresis in the discharge-sediment concentration relation (Figure 3.7). Sediment concentrations drop through the rainfall period. Klove suggests that this is not a sediment exhaustion effect, since sediment is transported in subsequent runs on the same plot; instead the study suggests that reduced sediment flux is a function of reduced particle detachment by raindrop impact as the surface wets up. Repeated storms on the same plots showed declining sediment yield suggestive of some measure of sediment exhaustion and Klove suggests that this is due to rill deepening exposing larger areas of unweathered peat surface. Comparison of runoff data at plot and channel scale demonstrates that the degree of inter-storm hysteresis in channels exceeds that from individual plots. Klove argues that this is because of the important role of channel floors in temporary storage of sediment which has been weathered from channel walls. A similar effect has been observed in naturally eroded peatlands in the Southern Pennines (Figure 3.8).

Overall the peat mine data point to the importance of weathering in sediment supply but indicates that timing of sediment delivery may be affected by rainsplash effects and will therefore vary with the nature of individual storms as well as with the antecedent weathering conditions. Two important caveats are required when attempting to relate Klove's findings to natural peat erosion. Firstly, the rainfall intensities used in the study ($35-240\,mm\,hr^{-1}$) are very high. Even the lower end of the range is only rarely achieved in summer convective storms in temperate latitudes. Secondly, the study focussed on peat mines so that the bare peat surfaces were significantly disturbed. Furthermore, Klove reports raking the surface between simulations so the results may not be typical of a naturally weathered bare peat surface.

A second rainfall simulation study carried out by Holden and Burt (2002c) more closely replicates natural peatland conditions with rainfall intensities of $3-12\,mm\,hr^{-1}$ applied to naturally eroded peat surfaces. The study also reports decreased sediment production as the storm progresses

Figure 3.8   Inter-storm storage of dry peat on the floor of eroded gullies in the Southern Pennines, UK) (photo reproduced by permission of James Rothwell)

and positive hysteresis in the runoff-sediment concentration relation. Holden and Burt suggest that rainsplash is an important mechanism in sediment supply but that sediment delivery is also affected by the transport capacity of micro-rills developed on the peat surface. Comparing spring and summer simulations, they argue that supply limitation of sediment flux is dominant in the spring but that in the summer the system may be transport limited.

In summary the limited experience of experimental approaches to investigating sediment production from bare peat surfaces confirms the suggestion from catchment scale studies that the role of peat weathering is important. However, whilst preparation of the surface is a necessary condition for erosion it may not be a sufficient one. The role of rainsplash appears important, either directly as a transport process or in mobilizing peat particles from the weathered surface. The process of sediment delivery from bare peat flats or from bare gully walls is not simple and the hysteresis observed in sediment concentration is due to more than simple exhaustion of pre-prepared sediment. In fact there appears to be a complex interaction of the antecedent state of the bare surface and the hillslope sediment transport processes operating within a storm event. There is also some evidence that the nature of this interaction may change over time as the nature of the weathered surface evolves.

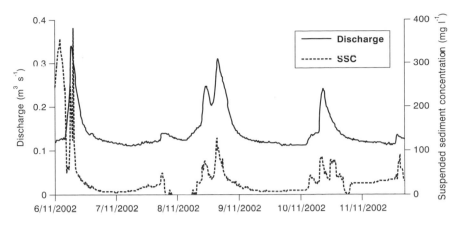

Figure 3.9   Example of inter-storm hysteresis. Suspended sediment concentrations for three storms calibrated from a continuous turbidity record. Upper North Grain, South Pennines, UK

3.5   Timescales of Sediment Supply

Sediment exhaustion has been identified as a feature of the peatland sediment supply regime at a range of timescales, including intra-storm (Burt and Gardiner 1984; Labadz et al. 1991; Yang 2005), inter-storm (Labadz et al. 1991; Yang 2005) and seasonal (Francis 1990). It is clear from the field and experimental evidence that sediment exhaustion simply conceived as the total removal of a weathered layer cannot account for the observed patterns in sediment flux at all of these timescales. Yang (2005) demonstrates a positive relationship between the degree of positive hysteresis in sediment concentration – discharge relations for a small eroding peat catchment and the degree of peat surface weathering measured at weekly intervals in sediment traps on gully walls. This is clear evidence of the importance of surface weathering at short timescales. It is however very rare to observe the complete removal of the friable layer and exposure of intact peat on bare surfaces during a storm. Similarly, although inter-storm exhaustion is commonly observed (e.g. Figure 3.9), significant sediment transport is observed to occur throughout series of storms. On the current evidence therefore the most likely mechanism to explain short time-scale sediment exhaustion is that identified by Klove (1998), i.e. the incision of micro-rills through the friable layer to the more resistant fresh peat below in combination with a reduced rainsplash contribution during the declining limb. Where a series of storms occurs in quick succession, the rill network may be re-activated with sediment supplied largely through rainsplash producing lower sediment concentrations and inter-storm hysteresis in the sediment concentration discharge curve. Longer intervals between storms allow time for weathering and frost action, in particular, produces rapid alteration to the surface microtopography such that subsequent rill erosion occurs in different locations increasing the initial sediment flux.

Seasonal patterns of sediment production with peaks in mid-winter and in the late summer (Francis 1990), and relatively lower sediment flux in the spring are more likely associated with changes in the depth of weathered sediment available at the surface controlled by the distribution of frost and periods of summer desiccation (Figure 3.5).

3.6   Conclusion

There is a considerable body of work on rates of sediment production from bare peat surfaces, most of it derived from erosion pin studies. Rates of erosion are relatively consistent across environments ranging from the sub-Antarctic Macquarie Island to the Shetland Islands. The measured surface retreat rates are rapid, ranging from $5–74\,mm\,a^{-1}$. What is less certain is the degree to which surface retreat rates equate to rates of sediment supply to the fluvial system. The available evidence suggests deflation and oxidation may account for a significant part of the total retreat, but both of these areas require further research (see Chapters 6 and 9).

Although the rates of hillslope erosion (and hence sediment supply to the fluvial system) are well known there are few data on the nature of the processes responsible. The general statement that sediment 'preparation' or weathering is an important precursor to erosion is well established. There is good evidence that frost action, particularly diurnal needle ice formation and desiccation, are the important weathering agents but their relative importance has not yet been clearly established, and probably varies geographically as a function of climate. Experimental work with rainfall simulators has shown that sediment supply is not simply a function of antecedent conditions and weathering but is also in part controlled by the nature of precipitation events and the role of rainsplash in sediment transfer. More detailed work is required to unravel the relative importance of these controls, and also to investigate the effect of variable peat composition. All of the preceding discussion has assumed that the peat is a homogeneous material, which is not the case. Peat profiles vary vertically reflecting past peat surface conditions and laterally reflecting local environments and previous erosion. Fibrous peat layers are commonly observed to coincide with breaks in the slope profile on gully walls due to their greater resistance to erosion but there has been no systematic research into the role of substrate quality on hillslope sediment production from bare peat.

This chapter has summarized the rather limited body of literature considering hillslope sediment production and presented some new data from sites in the Pennine Hills of England. It has suggested ways in which the existing understanding of the processes of sediment supply can account for empirical observations of the timing of sediment supply to the stream network. The nature of that stream network and sediment transport within the fluvial system is considered in more detail in the next chapter.

Chapter Four

Fluvial Processes and Peat Erosion

4.1 Introduction

Blanket peatlands are characteristic of hyper-oceanic climates and have characteristically low infiltration capacities, high runoff coefficients and flashy flow regimes. In such dynamic hydrological systems understanding of fluvial processes is central to an understanding of peatland sediment systems.

Most work on fluvial erosion of peatlands has focussed on the upper reaches of the fluvial system (e.g. Labadz et al. 1991; Carling 1983; Francis 1990), particularly on the zero-order hillslope gullies. Flow in these basins is typically ephemeral, yet due to the flashy nature of the discharge regime they contribute an important component of the overall discharge. This chapter is divided into three main sections. The first two consider organic sediment dynamics in hillslope gullies and main stem channels respectively. The third section considers the integration of these two subsystems focussing on linkage between slopes and channels and developing a conceptual model of the operation of fluvial sediment systems in eroding blanket peatlands.

4.2 Gully Erosion of Blanket Peat

Although the mechanisms for triggering peat erosion are many and varied, the controls on the physical removal of peat are more straightforward. Peat is removed from bare peat surfaces through the action of running water, wind and chemical oxidation. The action of running water is distinct from the other mechanisms in that it tends to be concentrated rather than diffuse. For this reason running water is the dominant mechanism in

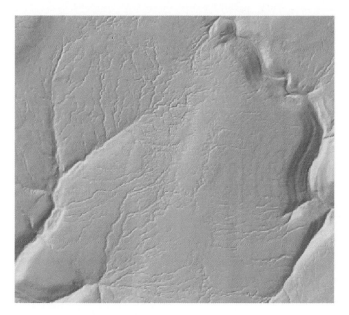

Figure 4.1   Dendritic channel network of an area of eroding blanket peat (3.1 km²) in the South Pennines. The image is a hill shaded DEM derived from 2-metre resolution LIDAR data

initial stripping of vegetation cover along drainage lines and the initiation of gully erosion. The morphology and topology of gully networks tends to support this view, with hillslope gullies showing connected, often dendritic, drainage patterns characteristic of fluvial systems (Figure 4.1).

4.2.1   Gully morphology and topology

The classic study of the morphology of eroding blanket peat was reported in a series of papers by Bower (1960a and b, 1961, 1962). This work involved exhaustive mapping of peat erosion in the Pennine hills of England, based on interpretation of aerial photography. Bower (1960a, 1961) made two key observations of eroding gullies in blanket peat. The first relates to the evolution of gully cross-sections. Bower identified four stages in the development of eroding hillslope gullies in peat (Figure 4.2). The transition from 'v' shaped gullies in the peat to a flat-floored 'u'-shaped profile is widely observed in eroding upland peats (e.g. Figure 4.3). Controls on re-vegetation of late-stage gullies are considered in detail in Chapter 8. Bower's second key observation refers to the plan-form pattern of eroding gully networks. Bower identified two dominant

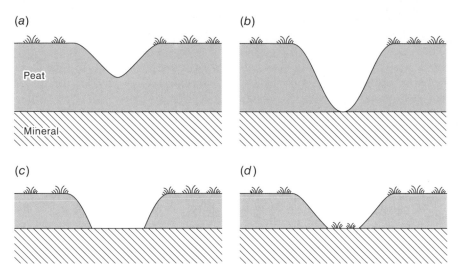

Figure 4.2   Four stages of evolution of hillslope gullies after Bower (1960a). (a) Initial 'V' shaped incision into peat. (b) 'V' shaped gully to full depth of the peat. (c) Development of flat floored profile as lateral erosion rates into peat exceed vertical incision into the mineral substrate. (d) Re-vegetation of eroded gullies

patterns of gully erosion which she termed Type I and Type II. Type I erosion is composed of frequently branching wandering channels, dendritic in form and with a high drainage density. Bower associated this type of erosion with lower slopes less than 5 degrees. In contrast Type II erosion is characterized by relatively straight unbranched channels aligned normal to the slope on steeper ground and with a lower drainage density (Figure 4.4).

The origin of hillslope gullies has been the subject of some debate. The association of the most intense Type I erosion pattern with areas of lower slope is apparently counter intuitive. Bower (1960b) suggests that the Type I erosion patterns develop from areas of hummock and hollow topography developed on relatively low gradients on the intact mire surface. Under this model connection of the pools by erosion of dividing hummocks produces the dense dendritic pattern. Support for this model is provided by Tallis (1994) who used palaeoecological evidence to demonstrate the probable existence of pool hummock complexes in South Pennine blanket peat prior to the onset of recent erosion. Tallis suggested that the gully erosion was initiated by erosion of hummocks, particularly associated with desiccation pressure on the surface vegetation during the Medieval Climatic Optimum (circa 1250 AD). These findings are consistent with the suggestion that vertical gully erosion was triggered by the

Figure 4.3   (a) 'V'-shaped gully profile in peat, Kinder Scout Plateau, South Pennines (b) Typical later stage, flat, mineral floored, gully, Bleaklow Plateau, South Pennines

crossing of an extrinsic geomorphological threshold as postulated earlier in Chapter 1.

The two further mechanisms of gully initiation proposed by Bower were headward incision of existing gullies and the collapse and consequent unroofing of large pipe systems within the peat. Gully erosion systems in both the North and South Pennines show clear evidence of frequent knick points within gully long profiles suggestive of headward

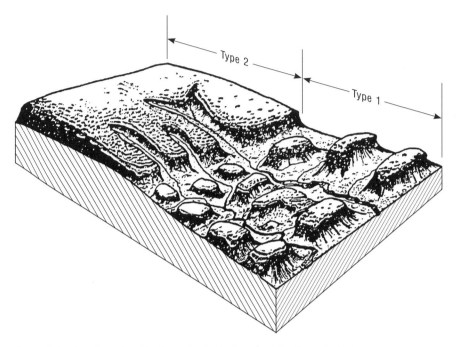

Figure 4.4   Type I and Type II gully erosion in blanket peat (after Bower 1960a)

migration of gullies. Bower (1960b: 329) noted that 'tributaries often hang above the main gully' and the discordance of tributary gully confluences is often striking. There has not been any detailed quantitative work to date on rates of gully headward migration but the morphological evidence in eroding systems suggests the process is active and relatively rapid.

Extensive development of soil pipes both near the surface and at the interface between peat and the mineral substrate is commonplace in blanket peatlands (Holden and Burt 2002a). Smaller gullies are commonly observed to disappear underground only to emerge a short distance downstream. Limited evidence of this mechanism can be observed on the summit of the Cheviot in Northern England where gullies develop from the collapse of large pipes formed at the peat–mineral interface. It is notable that the plan form of pipe networks reported by Holden et al. (2002) is qualitatively similar to Bowers Type II erosion form and it may be that pipe collapse is more commonly associated with this form of erosion on steeper slopes.

It is clear therefore that there is circumstantial evidence for a range of gully initiation mechanisms, but that there is relatively little hard evidence to rank the relative importance of these mechanisms. What is certain is that once gully systems begin to develop fluvial erosion of the peat is rapid, producing deep and extensive gully systems.

The Bower classification is an intuitively attractive system which subjectively appears to be a useful description of many eroded peat landscapes. Implicit in Bower's discussion of the classification is the suggestion that intermediate forms exist. Mosely (1972) used analysis of channel length and basin area, slope and width to demonstrate that eroding peat catchments, rather than exhibiting two modal topologies, show continuous variability and that the Bower 'erosion types' are characterizations of the end points of the distribution. Figure 4.5 exemplifies the continuous nature of erosion pattern under topographic control using data from Alport Moor in the Southern Pennines near to the site of Moseley's original work. Figure 4.5b shows a positive correlation between drainage density and mean slope. The data show considerable scatter but the correlation is strongly significant. The residuals from the regression line are normally distributed and the absence of grouping within the data supports the suggestion that slope is a continuous control on patterns of incision. The intercept of the regression line with the x axis suggests a threshold of 2–3 degrees of slope before significant drainage patterns develop which makes physical sense. The positive sign of the correlation appears to run counter to Bower's classification, but in fact reflects the fact that this analysis includes areas of uneroded peat which occur on hilltops at lower slopes. Talling and Sowter (1999), summarizing previous work from a wide range of environments, suggest that positive correlation between slope and drainage density is characteristic of landscapes where overland flow is dominant in drainage initiation. The observed pattern is therefore consistent with a dominant role for fluvial action in initiating peat erosion.

Bower's classification provides a useful shorthand for description of the form of peat erosion. Over half a century after its publication modern approaches to the analysis of landscape pattern (Gustafson 1998; Walsh et al. 1998; Belyea and Lancaster 2002), combined with new data sources such as laser altimetry (LiDAR) data (Figure 4.5) and the computational resources to manipulate the data, provide a significant opportunity to provide an enhanced quantitative understanding of erosion pattern, which is an important avenue for future research in this field.

4.2.2   Fluvial erosion in ephemeral hillslope gullies

Bower's work was of its time in that it inferred erosional process from qualitative observation of erosion form and pattern. The earliest process

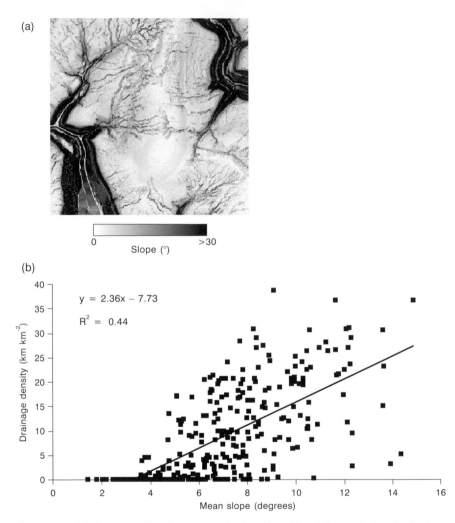

Figure 4.5   (a) Slope map of blanket peat moorland on Alport Moor (2 km × 2 km) in the Southern Pennines, UK. Derived from 2-metre ground resolution LiDAR data. (b) Correlation of mean slope with drainage density (calculated for 1 ha ground scale pixels in peat-covered area where this is taken as areas of slope <15 degrees) regression significant at 0.001 level

work on peat erosion was carried out by biologists working in the Pennines. Crisp (1966) made quasi-continuous measurements of sediment flux on Rough Sike in the North Pennines in order to produce a nutrient budget for the catchment. The data identified a typically flashy peatland hydrological regime with sediment flux dominated almost

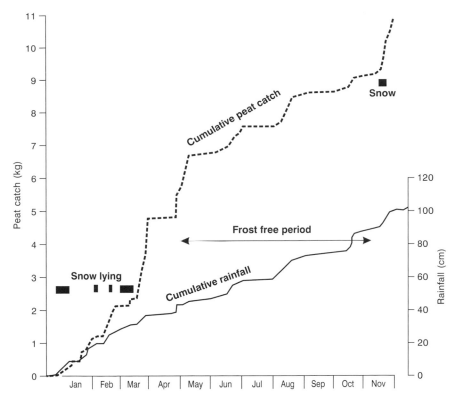

Figure 4.6 Cumulative sediment flux data from an eroding gully in the South Pennines illustrating reduced sediment flux during the frost free period (after Tallis 1973)

entirely by transport during storm events. It was estimated that 80–90 per cent of storm sediment flux occurred within two hours of the peak discharge. Combined with relatively high total sediment yields of 112 t km^{-2}a^{-1} and the largely organic nature of the sediment load these data were a strong indication that significant peat erosion in the catchment was fluvially driven. Another early quantitative study of the rates and processes of peat erosion was the work by Tallis (1973) attempting to measure erosion process in severely gullied terrain in the Southern Pennines. Using a large sediment trap, with a 0.9 millimetre mesh size which filtered a proportion of the stream flow, Tallis estimates the best estimate of total yield at 300–1,000 kg a^{-1} (equivalent to 12–40 t km^{-2} a^{-1}). Tallis's data demonstrated a clear association between rainfall- and snowmelt-generated runoff in the gully, and measured sediment flux. Cumulative sediment flux curves (Figure 4.6), show reduced rates of sediment flux during the frost-free summer months and the study concluded that winter frost action producing a friable surface layer on the bare gully walls was an important sediment supply mechanism.

Quantitative analysis of erosion processes from a geomorphological perspective was undertaken by two key studies in the late 1980s (Francis 1990; Labadz et al. 1991). Both Francis and Labadz worked on small catchments in eroding blanket peat, in Mid-Wales (Plynlimon) and the South Pennines respectively. These studies established several of the basic characteristics of the timing of sediment flux from eroding blanket peat catchments. As discussed in Chapter 3, stream sediment concentrations in relation to stream discharge show hysteresis at a range of timescales. Sediment concentrations normally peak ahead of stream discharge within a storm and during successive storms of similar magnitude peak concentrations typically decline (Labadz et al. 1991). Francis (1990) demonstrated that at the Plynlimon site hysteresis in stream discharge-sediment load relations calculated on a monthly basis also exhibit positive hysteresis with proportionally higher sediment loads in late summer and early winter and apparent sediment exhaustion in the late winter. Francis's data suggest that the dominant mechanism controlling sediment production is summer desiccation of peat surfaces and that maximum sediment flux in autumn/early winter reflects export of summer weathering of the surface peat. This finding is in contrast to much of the previous work on peat erosion (Bower 1962; Tallis 1973), and to Labadz's work, which suggested that winter frost action was the dominant weathering mechanism. It should be noted however that Francis's work was carried out during two very dry years (1983/4) and that the Plynlimon site is at relatively low elevation, and therefore has fewer frost days, compared with the Pennine sites which many of the earlier workers studied.

The control of sediment supply on sediment flux appears to be complex and temporally variable. However, despite some differences between the studies, Francis and Labadz both argue that sediment supply is an important control on the timing of peatland sediment flux and that weathering of bare peat surfaces prior to stormflow episodes is probably more important than direct fluvial gully erosion.

Yang (2005) has demonstrated quantitatively that the lag between peak sediment concentration and peak discharge, and also the gradient of best fit lines through measured discharge-sediment concentration relations for individual storm events, are controlled by sediment supply from eroding gully walls. The sediment production processes discussed in Chapter 3 are therefore the key to estimating and predicting sediment flux from small eroding peat catchments, with antecedent climatic conditions and the degree of pre-storm weathering playing an important role in determining the magnitude and timing of storm sediment fluxes.

4.2.3   Sediment delivery from hillslope gullies

The implication of the research to date on sediment flux in peatland streams is that these systems are weathering limited (Goudie 2004). These observations however refer largely to small gully erosion systems where peat eroded from gully walls is transferred directly to the channel. Geomorphologists working in a range of environments have increasingly demonstrated that sediment supply to stream systems is not only a function of primary sediment production but also of the efficiency of delivery of that sediment to the channel (Trimble 1981; Walling 1983; Phillips 1989; Marutani et al. 1999; Walling et al. 1999). The previous discussion has largely assumed that organic sediment delivery ratios (Walling 1983) in peatland systems are close to 100 per cent given the low density of organic sediment and the flashy nature of the stream discharge.

An assessment of the extent to which sediment supply to peatland streams might in part be constrained by the nature of sediment delivery requires a complete description of sediment production, transport, deposition and export from the catchment. Such a sediment budget (Swanson et al. 1982) requires identification and quantification of sediment transport processes, identification of linkages between these processes and storage elements within the catchment and quantification of these storage zones (Dietrich et al. 1982). The data demands of such an exercise require monitoring of a range of sediment fluxes over at least an annual cycle, and a mechanism for reasonable extrapolation of on-site results to the whole catchment. In areas of upland blanket peat which are typically isolated and subject to extreme weather the logistical difficulties of such an exercise are significant. Continuous monitoring of the complete range of sediment-transfer processes over representative periods is therefore not practical. An alternative approach is monitoring of a range of processes consecutively rather than concurrently with careful attention given to establishing the representativeness of the measurements and the degree of potential error. Essentially a synthetic annual sediment budget is created from a range of individual process measurements in a given catchment. Such an approach was adopted by Evans and Warburton (2005) and Evans et al. (2006) in order to produce organic sediment budgets for two contrasting blanket peat catchments in the Pennines.

Figure 4.7a illustrates the organic sediment budget for Upper North Grain in the Southern Pennines. This is a catchment affected by severe gully erosion with a relatively high annual organic suspended sediment yield of $195\,t\,km^{-2}\,a^{-1}$. The data appear to confirm that storage is an insignificant component of the sediment budget of actively eroding systems. Sediment production from eroding faces measured by erosion pins is

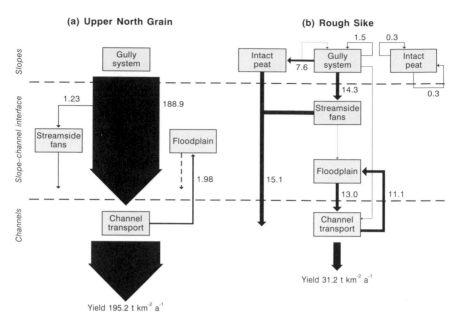

Figure 4.7   Organic sediment budget for two eroded blanket peat catchments (a) Upper North Grain, South Pennines. (b) Rough Sike, North Pennines

approximately balanced by measured sediment flux from the catchment. Storage of sediment by overbank deposition on the floodplain or by deposition at the interface of hillslope gullies and the floodplain is less than 2 per cent of the total sediment flux. Figure 4.7b shows equivalent data for an eroded blanket peat catchment in the North Pennines at Rough Sike. Total annual organic sediment yield in Rough Sike is substantially lower than the South Pennine site at $32 \, t \, km^{-2} \, a^{-1}$. The pattern of sediment production, transport and storage is also significantly different. Sediment transfer from the eroding gully walls to the stream channel is minimal as sediment from the hillslope gullies is deposited and trapped amongst vegetation at the downstream end of the gullies where vegetated fans separate the hillslopes from the channel system. Sediment supply to the channel in this system is largely derived from sources internal to the channel system being derived from either lateral floodplain erosion or localized sources where the channel cuts into the intact blanket peat. On Rough Sike therefore, significant re-vegetation of the floors of the hillslope gullies, particularly at their distal ends, has effectively decoupled the hillslope sediment system from the main channel. Sediment produced from the bare and eroding gully walls is stored on gully floors and at the vegetated interface between the slopes and the channel rather than being efficiently transferred to the channel.

The two sediment budgets described in Figure 4.7 therefore represent very different states of the eroded blanket-peat sediment system. Upper North Grain is an actively eroding system with limited re-vegetation of the hillslope gullies and efficient linkage of gully wall sediment sources to the main channel. Rough Sike is an eroded but partially re-vegetated system where sediment yields have been significantly reduced through re-vegetation of eroding gullies with limited linkage of the hillslope and channel sediment systems. At Rough Sike historical sediment yield data exists (Crisp 1966) which supports the inference of this transition indicating a reduction in sediment yield by a factor of 2.5 since 1962/3. The contemporary sediment yield appears to be a result of reduced slope–channel linkage associated with re-vegetation. Photographic evidence of local re-vegetation supports this hypothesis and clearly demonstrates that between 1958 and 1998 conditions have favoured natural re-vegetation of bare ground (Higgitt et al. 2001; Evans and Warburton 2005).

4.3   Erosion and Transport of Peat in Perennial Stream Channels

The discussion above has focussed on peat delivered to stream channels from eroding hillslope gullies. There is an implicit assumption in much of the writing on fluvial erosion of peat that once sediment is delivered to the stream channel it is rapidly and efficiently exported from the catchment due to its low density, but evidence increasingly suggests that the dynamics of both fine and coarse peat sediment transport within the channel system are complex. The role of main stream channels in eroding and transporting peat is considered below.

4.3.1   Production of peat blocks by fluvial erosion

Not all peat sediment transfer to the stream system occurs as fine grained detritus. Where peat banks are undercut directly by active stream systems, large blocks of peat may be deposited on the channel bed. The number of peat blocks transferred to stream channels by mass failure can be quite large, with some of the largest blocks having maximum dimensions in the order of metres. Experimental work has suggested that a critical submergence depth equivalent to the block depth is required to initiate peat block movement in streams (Evans and Warburton 2001); consequently in relatively small upland river systems only the largest flood events are capable of transporting failed peat blocks. The blocks are therefore stored in channel and on channel bars for considerable periods and have a significant effect on sedimentation within the channel. In larger peatland stream systems the major effect of peat blocks manifests itself on the plan form of the channel. Large blocks modify the flow and to a large extent control the deposition of gravels (Figure 4.8). Typically a tail of gravel deposition

Figure 4.8   (a) Peat block deposition in a gravel-bed stream in the North Pennines, UK. Flow is from bottom to top. Note the downstream tail of gravel deposition and significant scour around the block head. Also visible in the distance is the broken back of a large peat block, possibly caused by local scour in the adjacent pool. (b) A 1-metre-long axis peat block illustrating considerable effect on local gravel deposition patterns

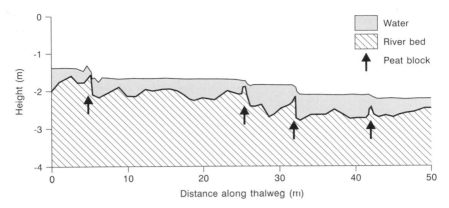

Figure 4.9 Stepped long profile along a 50-metre reach of Rough Sike, a small peatland stream in the North Pennines

is present downstream of the peat blocks but where multiple blocks occur large gravel structures are formed. Scour upstream of the blocks excavates a significant hollow around the block. Eventually undercutting of the upstream end of the block may cause cantilever failure of larger blocks producing characteristic broken blocks (Figure 4.8a) and accelerating the eventual block removal. In smaller stream systems the major effect of peat block inputs to the system is on the long profile of the stream. Large peat blocks form jams in narrow sections of the stream promoting upstream accumulation of bed material. In streams with active lateral erosion the effect may be to produce a stepped long profile (Figure 4.9).

The considerable effects of peat blocks on the form of stream channels and the interaction of the peat blocks with the mineral-bed material means that the eventual removal of peat blocks in large floods produces a significant reorganization of the stream system. Very little empirical work exists on the effects of peat block removal but the situation is in many ways analogous to the role played by large organic debris in forested stream systems (e.g. Gregory et al. 1985; Nakamura and Swanson 1993; Gurnell et al. 2002). In these systems woody debris dams locally impound the stream creating stepped profiles significantly lowering the channel gradient behind the dams. Periodic failure of debris dams in high flow introduces pulses of mobile mineral sediment into the system and these structural elements are a major control on bedload transport in many systems.

Two figures illustrate potential impacts of peat block removal. The first (Figure 4.10) illustrates changes in the plan-form of Trout Beck in the North Pennines during an extremely high-magnitude flood event in July

Figure 4.10   Eroding peat bluff on Trout Beck in the Northern Pennines. (a) Taken 22 September 2000. (b) Taken 21 August 2002, shortly after the major flood of 30 July 2002. Note the removal of the major peat blocks and then the post-flood reorganization of the channel showing the local influence of new blocks which fell from the bluff during the storm

2002. The flood had an estimated magnitude in excess of $30\,m^3\,s^{-1}$ which is the largest flood on record since records began in 1971. The large peat blocks present prior to the flood have been completely removed and there has been a dramatic reorganization of the reach particularly on the outer arch of the river bend. The new channel configuration is significantly modified by the presence of several large peat blocks which failed during the flood event. The degree of channel modification observed on this

Figure 4.11   (a) A peat block creating a step in the long profile of Rough Sike, a small North Pennine stream. (b) The same location after experimental removal of the block illustrating significant channel reorganization in response to block removal

reach of Trout Beck during the flood considerably exceeds that seen on other reaches where peat blocks are less important. This clearly indicates the importance of peat block dynamics in controlling the overall channel form. Figure 4.11 illustrates the effect of experimental removal of a peat block jam from a narrow tributary of Trout Beck. The removal of the block not only released quantities of stored bedload sediment from behind the jam but also triggered significant morphological adjustment as the channel widened adjacent to the block site.

4.3.2   Transport of peat blocks in stream channels

Once peat blocks begin to move they break down relatively rapidly through a combination of abrasion, splitting and slaking (Figure 4.12). Blocks roll initially and then float as the flow depth increases (Figure 4.13). Maximum abrasion rates occur whilst blocks are rolling (Evans and Warburton

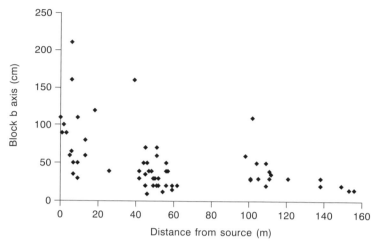

Figure 4.12   Decline in mean size of deposited peat blocks downstream of an eroding bluff, illustrating the rapid abrasion and breakdown of transported blocks. Data from Trout Beck, North Pennines, 2 February 1999

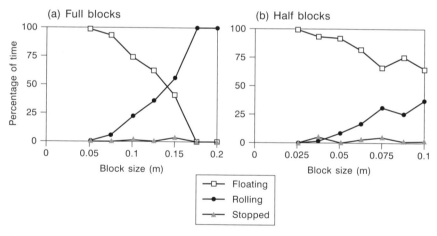

Figure 4.13   Relation determined experimentally between flow depth and peat block transport mechanism (after Evans and Warburton 2001)

2001), and specific abrasion rates are greater for larger blocks as they roll for longer periods during transport (Evans and Warburton 2001). There are two implications to the observed breakdown rates. Firstly, mean block size is reduced dramatically downstream. In Figure 4.12 mean block size 200 metres downstream of the peat block source is reduced to 20 centimetres. At this size the critical flow depth for block transport is reached during most storms in the main channel so the only remaining blocks stored in channel are those deposited high on bar-top surfaces during the receding flow of a flood. Therefore peat blocks only have a significant effect on channel plan-form (as described above) where the blocks are too large to be easily transported and where there is extensive lateral instability of the channel, and hence peat block production, at frequent intervals downstream. Secondly, the relatively rapid abrasion of transported blocks indicates that they constitute a significant in-channel source of fine sediment release, in addition to fine sediment delivered directly from the hillslope.

4.3.3   The fate of fine peat sediment in channels

Fine peat sediment is transported in suspension and in flotation in the stream channel. Despite the relatively low density of wet peat particles (close to $1\,g\,cm^{-3}$) not all suspended peat is transported. Where overbank flow occurs significant amounts of peat may be deposited on floodplains. Klove (1998) observed maximum rates of settling for saturated peat particles. One implication of this observation is that the proportion of peat deposited overbank may vary seasonally, with dry and hydrophobic peat released by summer desiccation being evacuated more efficiently than wet peat released by frost action. The presence of vegetation on the floodplain plays an important role in trapping peat particles. Newall and Hughes (1995) have shown that reduced flow velocities and lower turbulence are typical of areas of dense channel vegetation and Klove (1997) demonstrated maximum settling of peat particles in areas of shallow flow which had reduced turbulence. On the floodplains of many blanket peat systems in upland Britain, dense vertically growing grass, sedge, and rush species such as *Eriophorum vaginatum* and *Juncus squarrosus* are efficient at filtering particulate peat from overbank flows. Indirect evidence of storage of organic sediment within the channel system comes from work on fluvial carbon flux (Dawson et al. 2004). They demonstrate losses of particulate carbon load of 55 per cent between upstream and downstream sites in a peat catchment in northeast Scotland. Some of this loss is undoubtedly due to in-stream processing of the organic sediment and breakdown to produce dissolved organic carbon, but Dawson et al. (2004) also suggest in-channel storage of fine particulates as a possible cause. Evans and Warburton (2005) report overbank deposition equivalent to 36 per cent of the catchment sediment yield for a small North Pennine blanket peat catchment. Knowledge of the dynamics of floodplain deposition on peat is very limited. In this example the measured organic content of overbank

Table 4.1 Reported suspended sediment yields for upland peatland catchments. Values labelled (r) are derived from reservoir survey or coring. All other values from stream measurement of suspended load

Location	Catchment size (km²)	Sediment Yield t km⁻² a⁻¹		Reference
Wessenden Moor, S. Pennines	0.0042	55	Gully erosion of blanket peat	Labadz et al. 1991
South Pennines	Various	3–205 (r)	Range of Pennine reservoir catchments with many with significant blanket peat cover	White et al. 1996
Plynlimon, Wales	0.94	66	Shallow blanket peat cover	Francis 1990
Rough Sike 1960, N. Pennines	0.83	112	Eroded blanket peat (gully erosion and peat flats)	Crisp 1966
Rough Sike 2001, N. Pennines	0.83	44	Eroded and notably re-vegetated blanket peat	Evans and Warburton 2005
Upper North Grain, S. Pennines	0.4	265	Severe gully erosion of blanket peat	Evans et al. 2006
Gt. Eggleshope Beck, N. Pennines	11.68	12.1	Uneroded blanket peat moorland	Carling 1983
Gullet Syke	–	194	Eroding blanket peat	Wilkinson 1971
Langden Brook	–	232	Eroding blanket peat	–
West Grain, North Pennines	–	256	Eroding blanket peat	–
Hopes Reservoir, SE Scotland	7.0	26 (r)	Eroding moorland	Ledger et al. 1974
Monachyle, Balquhidder	7.7	38	Uneroded moorland	Stott 1997
Coalburn, Northumberland	3.1	3	Uneroded peat moorland	Robinson and Blyth 1982
Featherbed Moss, South Pennines	0.026	12–40	Eroding gully in blanket peat	Calculated from data in Tallis 1973
Burnhope Reservoir, North Pennines	17.8	33.3 (r)	Eroding peat moorland	Holliday 2003
Howden Reservoir, South Pennines	32.0	128	Eroding peat moorland	Hutchinson 1995

deposition was circa 75 per cent compared with typical floodplain sediment organic concentrations of 25 per cent. There are two possible explanations for this pattern; either the peat deposited overbank represents a flux rather than a sediment sink so that deposited peat is preferentially re-suspended and transported downstream in later storms, or alternatively there is significant loss of organic matter from floodplains by oxidation prior to burial. Further work is required to resolve these possibilities as the fate of floodplain-deposited peat has significant implications in terms of sediment flux, the fate of sediment associated contaminants (Rothwell et al. 2005) and the peatland carbon balance (Chapter 9).

4.4   Sediment Yield

4.4.1   Bedload yields

Bedload has only rarely been estimated for blanket peat catchments. Labadz et al. (1991) tabulated UK examples and illustrated that bedload as a proportion of total particulate load does not exceed 3 per cent. In systems where peat erosion is well advanced and sub-peat mineral sediment sources are well exposed the supply of bedload may increase beyond this level but the coarse sediment system is essentially separate from the delivery of peat sediment from hillslopes. The exception is where fluvial undercutting of the peat mass delivers bedload-grade peat boulders to the channel as discussed previously. However, the rapid abrasion of these clasts to fine sediment means that they are not a major component of total load.

4.4.2   Suspended sediment yields

Sediment yield data from entirely peatland catchments for comparison are relatively scarce. Table 4.1 reports published peatland suspended-sediment yields from the literature. These reveal a significant difference between sediment yields recorded for intact blanket peatlands and much higher sediment flux from eroding peatlands. The highest recorded sediment loads above $200\,t\,km^{-2}\,a^{-1}$. These values are relatively high within the overall UK upland context (typical values up to $100\,t\,km^{-2}\,a^{-1}$, Walling and Webb 1981), but the effective rate of land surface lowering that these yields represent is extremely high because of the low density of peat. Dry peat has a density of close to 0.1 so that volumetric removal of peat is on the order of 15 times greater than that represented by an equivalent dry mass of mineral material. The morphological impact of these high rates of removal is dramatic, producing the characteristic 'badland' landscapes (Tallis 1997a) of the UK's eroding blanket peatlands.

The onset of peat erosion is associated with the crossing of a geomorphological threshold and a very dramatic change in system state. This change of state is reflected in reported sediment yields. Intact blanket peatlands exhibit rather low suspended sediment loads. The dense moorland vegetation prevents significant hillslope erosion and under these circumstances the major control on sediment loads is the extent to which there is lateral erosion by stream systems into the peat mass. In contrast, once significant gully or sheet erosion of the peat-covered hillslopes is initiated very significant rates of erosion are possible. The considerable logistical demands of producing accurate sediment yield estimates for remote upland peat catchments explain the relative scarcity of reported work in Table 4.1. It is also obvious that work on peatland sediment flux, particularly from eroding catchments, has been largely a UK-focussed field of research reflecting the severely degraded nature of UK peatlands. It is therefore likely that the reported sediment loads do not span the full range of possible erosion rates.

An alternative approach to estimating the potential range of sediment loads from eroding catchments is to consider possible rates of sediment production since maximum long-term sediment yields should not exceed rates of sediment production in well-linked eroding gully systems. Sediment production from gullied catchments is controlled by the rate of retreat of bare peat gully walls and by the area of peat exposed in the gully walls. Maximum measured drainage density for the severely eroded South Pennine catchments presented in Figure 4.5b is $40\,km\,km^{-2}$. If this is taken as a maximum likely area-specific gully length for eroding catchments, and a typical mean peat depth of 2 metres is assumed, then maximum areas of exposed peat will be on the order of $80,000\,m^2\,km^{-2}$. Reported rates of gully wall retreat are summarized in Chapter 3 and a maximum likely retreat rate is on the order of $5\,mm\,a^{-1}$. This gives an annual volume of peat removal of $4,000\ m^3$ which, taking a typical density of $0.1\,t\,km^{-2}\,a^{-1}$ for upland peat, equates to a maximum sediment yield of $400\,t\,km^{-2}\,a^{-1}$. This is probably a maximum value for three main reasons: (1) it is unlikely that the maximum drainage density observed for a $100\,m^2$ square (as calculated for Figure 4.5) would be replicated across a whole catchment; (2) it assumes efficient sediment delivery and no storage; and (3) there are difficulties in interpreting gully wall retreat rates measured by erosion pins (Chapter 3). If in fact up to 50 per cent of retreat is shrinkage and oxidation then the estimates here of particulate sediment production should be reduced proportionately. Despite the considerable uncertainties associated with this approximation, maximum calculated values are of the same order as the largest yields shown in Table 4.1 which give some confidence that the relatively small sample of measured sediment yield describes most of the potential variation in the system.

4.4.3   Dissolved load

There are two main sources of data on dissolved loads in peatland streams. These are data from monitoring of total dissolved loads in streams, and data on dissolved organic carbon flux. Labadz et al. (1991) report total dissolved loads from a Southern Pennine catchment of $90–152\,t\,km^{-2}\,a^{-1}$. Total dissolved flux from the catchment is therefore up to three times the suspended sediment load ($55\,t\,km^{-2}\,a^{-1}$). These data derive from a peat-floored gully system so that the only possible sources of solutes are atmospheric deposition or solution of peat. Labadz suggests that around 20 per cent of the load may be accounted for by atmospheric sources. Where peat erosion has progressed to expose mineral sediment on the floors of gullies a further source of dissolved material is introduced. Worrall et al. (2003) and Daniels (2002) have demonstrated for sites in the North and South Pennines respectively that the hydrochemical system is dominated by two main sources of water namely surficial flow through the peat, rich in dissolved organic matter, and groundwater with higher concentrations of base cations. Temporal variability in solute flux appears to be controlled by simple mixing of the two water sources with peat waters dominating under stormflow conditions and higher groundwater contributions at low flow. In the South Pennines this can lead to very dramatic fluctuations in streamwater acidity dependent on the relative proportions of circum-neutral groundwater and flow through the peat with typical pH in the range 3 to 4.

Much more work has been carried out looking specifically at the organic component of the solutional load of peatland rivers. In particular a large number of studies have focussed on dissolved organic carbon (DOC) because of the potential importance of DOC flux in terrestrial carbon budgets (e.g. Chasar et al. 2000; Elder et al. 2000; Fraser et al. 2001; Glatzel et al. 2003; Worrall et al. 2003). Carbon is produced in peatlands in a soluble form through the microbial breakdown of plant materials to yield, in particular, humic and fulvic compounds. Hope et al. (1997) have demonstrated that in a series of catchments in northeast Scotland, the area of surface peat explains on average 65 per cent of the variance of DOC flux amongst catchments. Elder et al. (2000), however, have suggested that for catchments in northern Wisconsin the nature of hydrological connectivity within catchments as well as simple peat area is an important control. The key role of the exact balance of hydrological pathways has also been emphasized for Swedish Bogs by Laudon et al. (2004)

and by Fraser et al. (2001) working in Ontario, Canada. Recent work has also identified significant long term increases in DOC flux from peatlands (Freeman et al. 2001a; Worrall et al. 2004). Numerous studies have shown that the carbon dynamics of peatlands are strongly influenced by mean water-tables within the peat. Freeman et al. (2001b) argue that decomposition is inhibited by suppressed activity of a single enzyme, phenol oxidase, under anaerobic conditions beneath the water-table. This process is termed the 'enzyme latch mechanism'. Increases in DOC flux are implicated in many of the negative effects of peatland degradation including carbon loss, export of stored pollutant metals (Tipping et al. 2003), and colouration of drinking water (Naden and McDonald 1989; Pattinson et al. 1994). These impacts are covered in more detail in Chapter 9 but a detailed review of DOC work is beyond the scope of this volume. However, two important links between the fluvial erosion of peat considered in this chapter and changes in the dissolved carbon loads are examined here.

First, decreased water-tables in peatlands associated with increased drought frequency, possibly as a result of global climate change, have been implicated as possible causes of increasing DOC flux from intact peatlands. Where intense fluvial erosion of peatlands is occurring this effect may be exacerbated by local peat drainage. Where dense gully networks have developed, drainage adjacent to gullies can significantly reduce local water-tables. Where the impacts of artificial drainage on DOC flux have been monitored drainage has been shown to considerably increase streamwater DOC concentrations (Banas and Gos 2004). There is therefore potentially a complex positive feedback between increasing particulate sediment loads and organic solute flux from the system.

A second important linkage between dissolved and suspended loads in peatland streams is the potential for transformations between dissolved and particulate load. The literature on dissolved organic carbon in peatland streams focuses on generation of DOC in surface peats within the catchment and relatively little is known about in stream generation of DOC by dissolution of the particulate organic load. Dawson et al. (2004) describe relative proportions of dissolved and particulate organic carbon (particulate organic carbon loads are typically 50% by mass of the organic suspended sediment load) in a series of small catchments in intact blanket peatland in northeast Scotland. In these uneroded catchments the organic suspended load is very low, on the order of $3\,t\,km^{-2}\,a^{-1}$ (calculated from figures in Dawson et al. 2004, assuming POC content of 50%). The particulate load is 15–20 per cent of the dissolved flux demonstrating that in uneroded bogs the major mechanism of fluvial degradation of the bog is removal of material in solution. However, has noted previously, Dawson et al. also identify a 55 per cent loss of particulate load between upstream and

Upper North Grain	Rough Sike 1960	Rough Sike 2004
High sediment yield	⟶	Low sediment yield
Slope sediment sources dominate	⟶	Channel sediment sources dominate
Largely bare gully floors	⟶	Largely vegetated gully floors
High slope–channel linkage	⟶	Low slope–channel linkage

Figure 4.14   Conceptual model of eroding peatland dynamics identifying a spectrum of states from eroded through to re-vegetated. Associated changes in the sediment system are highlighted

downstream sites which is attributed to a combination of settling of particulates on the stream bed and in stream production of DOC from the particulate load. The relative importance of floodplain deposition of particulate peat identified for eroding blanket peatlands above suggests that floodplain storage may account for some of the observed transmission losses but it seems likely that some degree of dissolution does occur. Koelmans and Prevo (2003) demonstrated experimentally that turbulent mixing of organic matter in a water column promotes relatively rapid breakdown of particulates, and it is possible that in the more turbid waters of eroding catchments this mechanism may be a significant source of DOC production. The role of erosion and particulate carbon flux in the wider carbon budget of blanket peatlands is discussed in more detail in Chapter 9.

4.4.4   A conceptual model of sediment dynamics in eroding blanket peatlands

The evidence from the two sediment budgets described in Figure 4.7 together with the suggestion that the Rough Sike catchment has seen dramatic reductions in sediment yields over the last 40 years suggest that even within the class of eroding blanket peatlands there is considerable variability in sediment dynamics and consequently variation in sediment yields. Figure 4.14 is a conceptual model which describes the

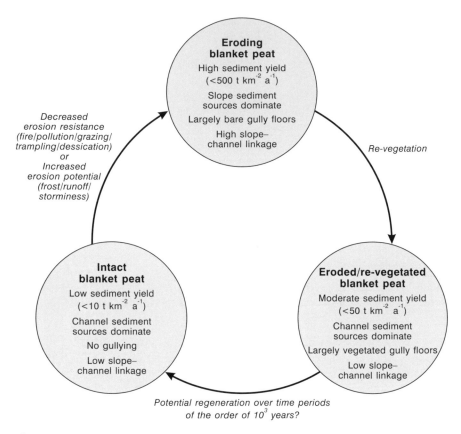

Figure 4.15   Peat erosion status and changes in fluvial sediment dynamics

characteristics of peatland sediment dynamics from uneroded blanket peat to severely eroded systems. The model suggests that the onset of erosion in blanket peat catchments is associated with increases in the sediment yield of up to two orders of magnitude due to massive extension of the fluvial system as hillslope gullies extend, sediment supply from bare peat is increased, and the degree of coupling between the hillslopes and the channel rapidly increases. If conditions permit re-colonization of the eroded system, re-vegetation of gully floors and mouths rapidly reduces sediment supply to the channel. The degree of linkage between slopes and channels is significantly reduced and lower sediment yields are increasingly dominated by channel sediment sources rather than delivery of eroded peat from the hillslope gullies. The model is also presented in a cyclic form (Figure 4.15), as over time continuing re-vegetation and renewal of peat growth on the gully floors can potentially regenerate a peatland system more akin to intact mire, although the timescales are likely to be long. The possibility of cyclic erosion and regeneration of blanket bogs is considered in more detail in Chapter 8.

4.4.5 Sediment yield, sediment supply and assessing catchment erosion status

Sediment yield, although a crude estimate of sediment dynamics, is still the critical criterion in assessing the erosion status of blanket peat catchments. Ecological assessment of bog status often includes reference to erosion status but typically this is derived from the extent of bare peat in the catchment, or the morphological expression of gully erosion. The discussion of sediment delivery and re-vegetation above suggests that such parameters are at best indirect indicators of contemporary erosion. Measurement of sediment yield provides not only an assessment of current erosion status but an indication of the rate of potential future degradation. As noted previously, estimation of annual sediment yields is logistically demanding, requiring repeat sampling over considerable periods. It would therefore be of practical value to develop a method for estimating sediment yields based on more limited field investigations. One possible approach is to take advantage of the strong sediment supply control on peatland sediment yields so that in regions of reasonably constant runoff regime catchment characteristics may control sediment yield. Yang (2005) has demonstrated that the slope of sediment concentration–stream discharge relations (the sediment rating curve) is highly sensitive to catchment sediment production driven by synoptic scale climatic variability. Due to the high degree of variability in sediment yields between catchments across the spectrum of erosion conditions, variability between catchments is likely to significantly exceed inter-annual variability at a site. It is possible therefore that the larger-scale control on sediment supply exerted by changes in sediment delivery and connectivity between catchments with varying erosion status is also reflected in the form of the rating curve. Figure 4.16 presents some preliminary data to support this hypothesis. The figure plots four suspended sediment rating curves from two eroding blanket peat catchments in the Pennines. The curves represent systems spanning the spectrum of eroding mire types. Two curves are presented for the severely eroded Upper North Grain catchment in the Southern Pennines (sediment yield $265 \, t \, km^{-2} \, a^{-1}$). One is the general rating curve for the catchment and the other a rating curve for a three-month period following a 100-year-return-period flood event in July 2002. The lower slope of the second curve is due to reduced sediment supply

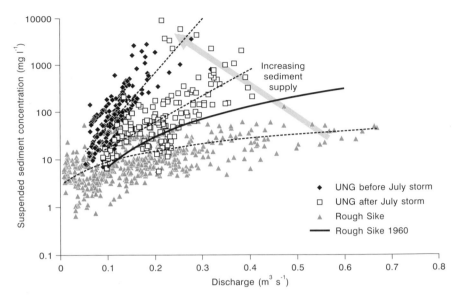

Figure 4.16   Illustration of variations in the slope of the sediment rating curves for two Pennine stream systems under conditions of varying sediment supply

due to sediment exhaustion. The major storm event was observed to strip the bare peat gully walls of much of their surficial layer of weathered peat. The lowest slope of the four curves is a contemporary sediment rating curve from Rough Sike (sediment yield $44\,t\,km^{-2}\,a^{-1}$). The final curve is that reported by Crisp (1966) for Rough Sike in the late 1950s when sediment yield was $112\,t\,km^{-2}\,a^{-1}$.

Because of the storm-dominated nature of sediment flux from peatland catchments most reported blanket peat catchments have low-flow suspended-sediment concentrations in the range 0–$10\,mg\,l^{-1}$. Sediment rating curves for stream systems across the spectrum of erosion types therefore converge on the origin. The rate of increase of sediment concentration with increasing discharge is therefore the major difference between curves for different catchments and might reasonably be regarded as a measure of the catchment's sediment supply status. Within a given climatic region where specific discharge from different blanket peat catchments is reasonably invariant the slope of the curve might also be a useful proxy for sediment yield. This potentially useful relation needs further testing with suspended sediment datasets from a wider range of catchments.

4.5   Conclusions

Fluvial erosion processes are the dominant mechanism controlling sediment flux from eroding peatland systems. Sediment yields from eroding peatlands measured in dry mass are modest in global terms but reflect very considerable morphological change in eroding catchments because of the low density of the eroded material. Sediment flux is significantly constrained by sediment supply at two distinct scales. Firstly, at the scale of ephemeral headwater gullies, weathering controls the generation of friable weathered peat from bare gully walls and peat flats limit sediment supply to the channel system. Secondly, at a larger scale, sediment delivery from the ephemeral gullies to the perennial stream network is controlled by the efficiency of sediment transfer from the slope system to the main channel. It has been argued here that the degree of re-vegetation of the eroded system is a key control on slope-channel linkage. In systems where re-vegetation is extensive a large part of the gully network can become effectively decoupled. In this situation sediment flux is controlled by in-channel processes, in particular by erosion of floodplain sediment and direct erosion of the peat mass. Despite the low density of eroded peat significant re-deposition occurs within the upland peatland catchments. Eroding peatlands are therefore highly dynamic systems with characteristic changes in the nature of the sediment budget as catchments erode and then re-vegetate. The sediment budget framework, with its emphasis on sediment supply and sediment transfer, is therefore an essential methodological tool for proper consideration of the fluvial geomorphology of peatland systems.

Chapter Five

Slope Processes and Mass Movements

5.1   Introduction

The mass movement of peat on slopes, commonly reported as peat slides and bog bursts, has been well documented for several centuries (Honohane 1697; Ouseley 1788; Griffiths 1821; Bailey 1879; Standen 1897; Crofton 1902; Feehan and O'Donovan 1996); yet the fundamental controls of this form of shallow instability are still relatively poorly understood (Tallis 2001). Although the majority of reported peat mass movements are restricted to the British Isles (Bowes 1960; Crisp et al. 1964; Tomlinson and Gardiner 1982; Alexander et al. 1986; Mills 2002), numerous examples occur throughout the world including: Germany (Klinge 1892; Vidal 1966); Switzerland (Feldmeyer-Christe and Mulhauser 1994; Feldmeyer-Christe 1995); Canada (Hungr and Evans 1985); Argentina (Gallart et al. 1994); Nyika Plateau, Malawi (Shroder 1976); Wingecarribee Swamp, Australia (Tranter 1999); and the sub-Antarctic Islands (Selkirk 1996; Nel et al. 2003).

Recently there has been a resurgence of interest in peat mass movements due to the occurrence of several catastrophic peat-slide events in the UK and Ireland (Figure 5.1). These include the multiple peat slides and landslips on Dooncarton, County Mayo, Western Ireland, on 19th September 2003 (Tobin, 2003; Dugan, 2004); the Channerwick, South Shetland peat slides also of 19th September 2003 and the Derrybrien bog failure in County Galway on 17th October 2003. These events have coincided with new research which aims to examine the fundamental characteristics of peat mass movement events and move away from the case-by-case approach which has dominated the literature to date (Mills 2002; Kirk 2001). This new approach is best demonstrated in several recent review papers which have attempted to produce general summaries

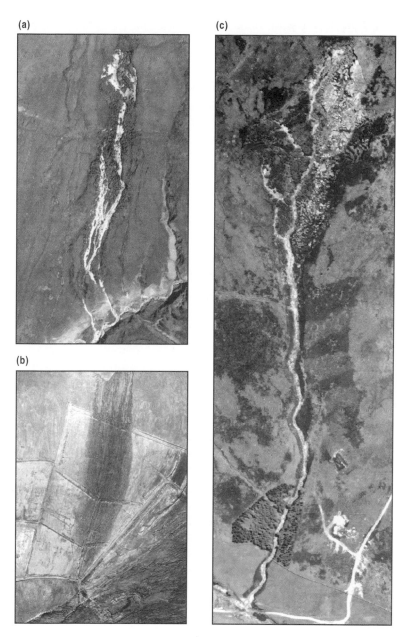

Figure 5.1 Examples of peat mass movements. (a) Harthope, North Pennines, Northern England, 1995. (b) Dooncarton, Co. Mayo, Ireland, 19 September 2003. (c) Bilsdale Moor, North York Moors, Northern England, 19 June 2005. Vertical scales are: (a) 600 metres, (b) 750 metres and (c) 950 metres

of the nature and controlling mechanisms of peat mass movements (Warburton et al. 2004; Dykes and Kirk 2006; Jennings 2006).

In many respects peat hillslopes can be viewed like any hillslope; and in this sense our basic understanding of their hydrology, soil mechanics and slope stability still applies (Selby 1993; Crozier 1986). However, the special properties of peat, particularly its hydraulic behaviour and unusual geotechnical properties (Boelter 1968), mean that the processes operating can be significantly different (Warburton et al. 2004). An early account by Crampton in 1911 recognizes the special features of mass movements in peat. Crampton (1911: 51) states:

> A point of some importance is the existence of movements in the peat. Great bog slides endangering life and property, such as occur from time to time in Ireland, are unknown in Caithness, but there is abundant evidence that movements of a less violent nature occur. The corrie on the south side of Beinn nam Bad Mhor is full of plastic peat which is gradually over-flowing and sliding down the flank of the mountain. In less plastic peat large cracks are of common occurrence, and often serve as a passage for water from above or below. The cracks seldom gape, and are often difficult to detect but for a change in the plant association in their vicinity. The stems of *Juncus effusus* may sometimes be seen growing in single file along cracks in peat covered by *Calluna* moor. More or less horizontal slickensided surfaces may sometimes be observed at the junction of the peat with the underlying drift, and also between different layers of the peat itself. The direction of the grooves is that of the local slope of the ground. Movements in the peat due to gravity would have as predisposing causes slope of the ground, thickness of the peat, and super-saturation of the peat and underlying deposits. Extensive cracks and areas of subsidence are often produced by the undermining of the peat by water erosion where the underlying deposits are of a loose nature. Cracks and slight movements are also produced through shrinkage of the peat by drying.

More recently, the Department of the Environment (1994) in their review of landsliding in Great Britain suggest organic materials are subject to failure like any ground forming materials, but the nature of the materials determines that the characteristic mass movements are slightly different. This review recognized three types of failure: bog bursts, bog flows and peat slides. However, in the peat mass movement literature the exact definition of these particular mass movements is uncertain and at times contradictory. In the general landslide literature the only classification schemes that explicitly mention peat mass movements are Crozier (1986) and Hutchinson (1988). In Hutchinson's scheme, peat mass movements are classified under both landslides (translational slides – peat slides) and debris moving as a flow-like form (debris flows, involving peat; bog flows,

Table 5.1 Environmental impacts of rapid peat mass movements

	Short-term impact	Long-term impact
At the site of the event	• Loss of farm stock and a danger to wildlife • Loss of agricultural standing crop • Displacement of peat mass and disruption of hydrological continuity	• Loss of an important ecological and palaeoecological resource • Site of continued erosion • Alteration of natural drainage • Loss of terrestrial stored carbon
Offsite effects	• Damage to infrastructure and housing • Damage to channel ecosystems, e.g. fish kill • Contamination of upland reservoirs and lakes	• Increase in lake and reservoir sedimentation rates

bog bursts). For the purposes of this chapter a simple approach is adopted which recognizes the significance of three dominant processes on peat hillslopes, namely: flow, sliding and creep. Slow mass movements dominantly involve creep whilst rapid mass movements are dominated by rapid flowing and sliding mechanisms (Statham 1977).

Peat slides and bog bursts are important because they mobilize considerable quantities of surficial peat, have major impacts on stream ecosystems (McCahon et al. 1987), and are a significant hazard in upland peat environments (Colhoun et al. 1965; Creighton 2004; Jennings 2006). For example, a slide in Baldersdale in northern England in 1689 caused considerable flood damage and poisoned fish for several kilometres downstream (Archer 1992). More recently the peat mass movements at Dooncarton Mountain in Western Ireland have cost the region in excess of $10 million (Dykes and Kirk 2006). The impacts of rapid peat mass movements are felt at-a-site, both in the long and short term, but also have wider off-site implications (Table 5.1).

The objectives of this chapter are to discuss the stability of peat covered hillslopes, describe the main morphological features of peat mass movements and discuss the various factors and mechanisms which lead to peat slope instability. Although this chapter has the broad title of 'slope processes and mass movements' the scope of the discussion is limited mainly to rapid mass movements involving peat slides and bog bursts. This is justified on the grounds that fluvial processes, involving gully erosion on slopes and wash processes, are covered elsewhere in this volume (Chapters 3 and 4) and little is known about slow mass movements in peat, even though the use of terms such as creep and tearing are frequently applied in the literature to describe such processes (Pearsall 1950; Bower 1960; Tallis 1985a). Details regarding the linkages between peat failures and hydrology are covered in detail in Warburton et al. (2004), therefore treatment here is limited to an overview.

5.2   Peat-Covered Hillslopes

A limiting factor governing the formation of thick peat deposits is local topography (Tallis 1985a). Peat tends to be thickest in closed depressions and, in areas of blanket peat, thins as the local slope angle increases (Campbell 1981). On steep slopes, thick contiguous peat deposits are unlikely to be found because drainage conditions are such that peat cannot form. Tallis (1985a) has used the term 'slumped peat' to describe where the peat mass has disintegrated into a series of blocks at the margin of the peat blanket. Therefore when considering slope processes and mass movements it is important to recognize that there is a limiting range of slopes over which these processes occur and this tends to be biased to lower angled slopes generally less than 20° (Bower 1960).

5.2.1   Limits to the stability of peat on slopes

The idea of a limiting angle for the occurrence of peat deposits can be traced back to a number of early writings (e.g. Crampton 1911), but its origin is perhaps most closely associated with Pearsall (1950: 272–4) in his discussion of peat erosion. Pearsall argued that on steep upland slopes peat becomes unstable when a certain critical depth is reached. He argued that the form of the peat slope will respond to what Pearsall termed the 'fluidity' of the peat. Pearsall (1956) developed these arguments further by suggesting that a peat bog on a slope could be represented as a fluid droplet bounded by a viscous skin of vegetation. Figure 5.2a is reproduced from Pearsall (1956) and compares the idealized bog profile with a profile more characteristic of that observed in the field. Peat depths tend to be relatively uniform through the profile. This implies that the peat is cohesive and well attached to the substratum and the bog systems are actually relatively stable on the low-angle slopes on which they occur. However, Pearsall suggested that desiccation of the bog would result in surface deformation due to loss of mass (water). Dependent on the basal slip condition between the peat and substratum two possible forms could develop. Firstly, in a bog where there is no basal slip the surface

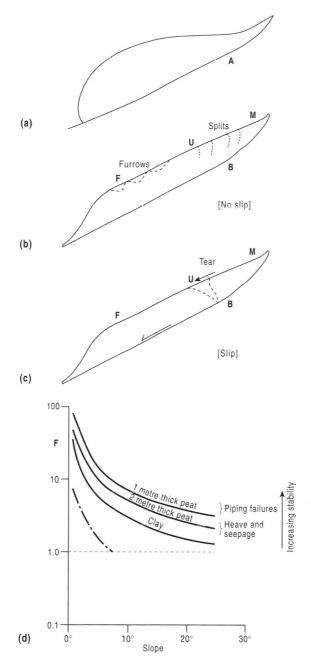

Figure 5.2 Schematic representation of hypothetical peat bog profiles on sloping terrain (from Pearsall 1956 [a–c]). Part (d) shows theoretical stability curves for Noon Hill peat and clay (from Carling 1986a). See text for further discussion

vegetation skin would be lowered and furrowed, resulting in the splitting of the surface peat (Figure 5.2b). Alternatively under conditions of basal slip, peat at U would be dragged downslope to F and a large tear might result between U and M. The toe of the bog, downslope of F, would remain wet and less susceptible to desiccation (Figure 5.2c). Under these conditions peat would continue to grow until it became deep enough to become unstable and start to erode.

Pearsall (1950, 1956), Bower (1959, 1961) and Johnson and Dunham (1963) all regard hillslope peat as inherently unstable, but evidence for such a relationship was not properly quantified until Carling (1986a) examined the stability characteristics of saturated hillslope peat. Based on the observation that peat thickness is inversely related to slope angle it is implied there may be a maximum depth for long-term stability of peat. Carling (1986a) produced critical-state intrinsic instability curves for a saturated hill peat and calculated these for two peat thicknesses of approximately 1 and 2 metres respectively (Figure 5.2d). It was demonstrated that the peat was mechanically stable over a range of slope angles in excess of the maximum slope angle on which peat deposits occur. This conclusion suggests for this particular set of circumstances the processes controlling peat formation are the primary factors governing the occurrence of peat on slopes in this environment. The material strength properties of the peat mass are therefore of secondary importance in determining the distribution of peat, except perhaps on a very local scale (Bower 1960; Campbell 1981).

This simple example is useful because it highlights several important factors which govern peatland slope processes. These are:

1 slope topography and form;
2 hillslope hydrology and hydraulic properties of the peat;
3 peat material properties.

Another aspect of Figure 5.2d which is relevant to the discussion here is the significance of the underlying (non-peat) hillslope materials in governing the overall stability of peat-covered hillslopes. In the example quoted in Carling (1986a; Noon Hill, northern England), it was demonstrated that failure in a horizon within clay a few centimetres below the base of the peat was the main cause of hillslope failure. This raises some interesting questions about what defines a peat mass movement. For the purposes of this chapter a peat mass movement is defined as any downslope movement of soil material under the influence of gravity where peat constitutes the main mass of displaced soil or where the distinct material properties of peat can be directly related to the mechanism of movement. For example, organic soil water leaching peat may be responsible for remoulding of sensitive clays at depth in the soil profile (Söderblom 1974) or clay disaggregation (Caillier and Visser 1988). However, a thin peat rafted over a deep-seated failure in clay cannot be considered a peat mass movement because peat is not a primary factor influencing the process, nor the dominant material in the failed mass.

5.2.2   Creep on peat hillslopes

Rapid mass movements in peat on short timescales of seconds to hours have dominated the literature on peatland slope processes (Warburton et al. 2003; Dykes and Kirk 2006). However, peat soils are also characterized by their high compressibility and long-term settlement, making them susceptible to slow en-masse movement over much longer timescales of years and decades (Barden and Perry 1968). These properties mean that, following primary consolidation, significant peat creep can occur. The rate of creep is determined by effective stress, peat voids ratio and soil fabric (Edil et al. 1994). In the field context, creep can be defined generally as the gradual downslope movement of peat on a hillslope under the influence of gravity. However, in an engineering sense creep is the slow time-dependent strain in a solid material. Given variations in soil material properties different forms of creep can be defined depending on how a sample yields under stress (Selby 1993).

Measurements of soil creep in the geomorphic literature are rather limited, particularly on organic soils and peat. An exception is work undertaken by Anderson (1977), working in Upper Weardale in the Northern Pennines. Between 1972 and 1975, Anderson undertook extensive soil-creep measurements using six different techniques in a small experimental catchment in Rookhope. The measurement period was 28 months, with detailed information collected from 20 sites on a relatively poorly drained, gently sloping (11°) valley-side slope primarily vegetated with *Nardus*, *Juncus*, *Pteridium* and *Calluna*. Peaty soils and peaty gleys occurred in the wetter areas of the slope at depths of up to 0.8 metres. Results are summarized in Figure 5.3, which shows the importance of organic content in relation to creep rate. Anderson reports a correlation of 0.81 between the two variables, although the unusual distribution of organic contents in the soil should be noted when examining the trend. Figure 5.3 also demonstrates that peat soils form a distinct group from the upland soils and creep rates tend to be twice as great on peat soils as their less organic counterparts. It is therefore questionable whether there is a simple relationship between organic content and creep rate. Given the distinctive, geotechnical and hydrological characteristics of peat such a distinction is not surprising.

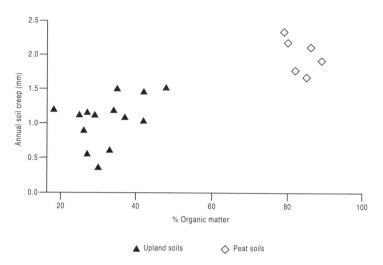

Figure 5.3   Measurements of soil creep from Upper Weardale in the Northern Pennines, UK, 1972 to 1975 (Anderson 1977). The measurement period was 28 months, with detailed measurements at 20 sites, six in peat soils

5.3   Morphology of Rapid Peat Mass Movements

Peat slides and bog bursts are characteristic rapid mass movements of peat hillslopes (Figure 5.1). A basic distinction is often made between two main types of failure: slides and bursts. Peat slides are generally best described as shallow translational failures, with a shear plane located at or just below the peat-substrate interface/transition (Crisp et al. 1964; Carling 1986a; Selkirk 1996). Bog bursts, in contrast, involve the rupture of the peat surface or margin due to subsurface flow or swelling. Liquefied peat is often expelled from peat faces or through surface tears. This results in fracturing and tearing of the hillslope, as the overlying peat mass is deformed following evacuation of the failed material (Hemingway and Sledge 1945; Bowes 1960; Tomlinson 1981a). Hobbs (1986) suggests that peat bursts involve much more fluid peat than do slides. Following failure, both bursts and slides will runout downslope and may rapidly degenerate into a flow of well-lubricated blocks, leading to their description as peat and bog flows (Colhoun et al. 1965; Alexander et al. 1986). Through time these sites will re-vegetate and leave only subtle scars in the landscape (Feldmeyer-Christe and Küchler 2002; Mills 2002).

Although differences exist between specific types of peat mass movement there are four main morphological elements common to most settings (Figure 5.4 and Figure 5.5):

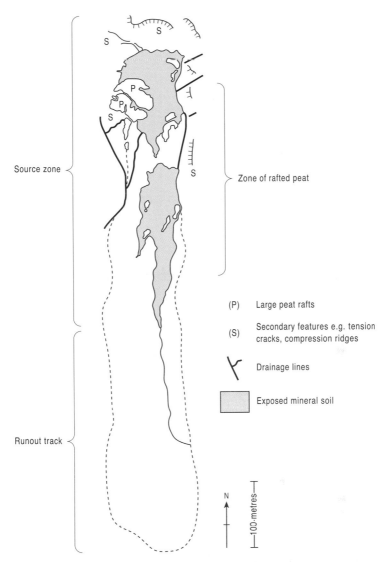

Figure 5.4 Nein Head II peat slide, North Pennines, UK, showing the four basic morphological elements that can be commonly identified in peat mass movements (based on Mills 2002)

Figure 5.5 Morphology of peat mass movements, Channerwick peat slides, South Shetland, September 2003. (a) Source zone. (b) Zone of rafted peat debris. (c) Runout track. (d) Secondary feature – extension tension cracks. (e) Secondary feature – rupture tension cracks. (f) Secondary Feature – compression ridge with ridge-crest cracking

1   a source zone consisting of single or multiple, often crescentic, scar areas (defined by an upslope head scarp);
2   a zone of peat debris dominated by large rafts of peat, and smaller blocks arranged in blockfields and clusters;
3   a runout track with abundant abraded and fractured peat blocks and a trail of peat slurry and uprooted vegetation. This is often bounded by distinct levees of peat blocks;
4   secondary features adjacent to the main failure showing evidence of both tension and compression within the peat, including both extension and rupture tension cracks; and compression ridges.

These basic morphological elements are common to many types of shallow mass movement (Crozier 1986). However, given the low density of peat and its unique yet highly variable geotechnical and hydrological properties, differences exist between peat mass movements and those in predominantly mineral materials. The four morphological elements described above tend to form overlapping zones within a peat-mass failure (e.g. Figure 5.4). These are now illustrated in relation to the example from the September 2003 peat slides in the Channerwick area of South Shetland shown in Figure 5.5.

5.3.1   Source zone

This generally consists of a single or multiple scar area which often has a crescentic head scarp (Figure 5.5a). The scars can be swept clean of all surficial material or may contain remnant peat rafts, blocks, slurry or peat boulders. The condition of the scar surface in the immediate aftermath of the failure can provide valuable information on the mechanism of instability. For example there may be a clear definition of a break between the overlying peat and substrate indicating failure at the interface between the two materials. This may often be accompanied by slickensides and deep gouges in the surface of the substrate where boulders and wood imbedded in the base of the peat have eroded the surface as the overlying mass has slid over. Alternatively a smear or veneer of peaty material may be left on the failure surface suggesting failure within the overlying peat. In some bog bursts the scar surface may not be exposed and the failure area is completely composed of peat of considerable depth indicating failure within the peat profile well above the mineral substrate. Alternatively failure may be at some depth into the mineral substrate and, although this may appear similar to a failure at the peat/mineral interface, small steps may occur in the scar area at the margins and the failure plane can be traced within the clay underlying the peat. Often the failure plane is not consistent and may migrate between the mineral layer, interface and overlying peat. A common occurrence of this is where the slope has failed at some point downslope from the head scar in the mineral substrate and this has resulted in unloading of the slope and retrogressive failure of the peat upslope. In this case the failure plane may have migrated from the mineral soil to the peat mineral/interface as blocks have slid downslope. The correct interpretation of such a failure requires a full survey of the scar area soon after failure. Post-slide degradation of the failure surface is often very rapid, therefore evidence is often masked very quickly (Praeger 1906; Large 1991; Mills 2002).

5.3.2   Rafted peat debris

The peat debris released from the source zone is displaced downslope over distances from a few centimetres to hundreds of metres. As the debris moves, it fragments, and the nature of this fragmentation process depends on the material properties of the peat overburden, the dynamics of movement and the roughness and local topography of the underlying failure surface. Initially large rafts of peat are detached from upslope and as they move downslope they tend to disaggregate into smaller peat blocks (Figure 5.5b). Peat rafts are generally considered to be large parts of the peat blanket whose planform dimensions are significantly greater than their depth such that the rafts move purely by sliding, and can form thrust features as they move downslope. Peat blocks are smaller masses of peat whose three principal axes are approximately equal and therefore may move by a combination of sliding and rolling. In fibrous peat, which generally has higher tensile strength, the size of the rafts and blocks is often proportional to the average peat depth in the failure zone, but also depends on the local peat stratigraphy and geotechnical properties (Warburton et al. 2003). Inspection of the base of the blocks provides important evidence for the location of the failure plane at the time of failure. For example in Figure 5.5c the blocks in the foreground of the photograph show a thin layer of mineral soil attached to the peat base. This is a clear indication that failure occurred within the soil substrate and not the peat.

5.3.3   Runout track

Continued movement of the disaggregated peat mass downslope usually results in the breakdown of the peat by mechanical abrasion, splitting and rolling (Colhoun et al. 1965). The debris consists of abundant abraded

and fractured peat blocks and a trail of peat slurry and uprooted vegetation (Figure 5.5c) (Latimer 1897). The resultant peat fragments often
form small peat peds with a spindle form. At the same time much of the
abraded material mixes with local water sources (both runoff and stored
water) forming a peaty slurry. The composition of the slurry and its rheological properties depend on the nature of the peat source, its initial condition (wet or dry) and the relative proportions of the peat/water mix
(Mills 2002). The runout track is often bounded by distinct levees or
lobes of peat blocks (Figure 5.5c). If the flow becomes confined to a
channel and there is evidence of super-elevation at bends in the flow track,
peat levees can be used to reconstruct flow velocities. For example, the
preservation of peat debris along Hart Hope Beck following a peat slide
in the North Pennines (Warburton et al. 2003) made it possible to estimate the flow velocity of the peat mass movement as it entered a valley
and moved downstream. Assuming the runout was moving as a debris
flow the entrance velocity into the valley and downstream velocity can be
approximated using the velocity head equation (Hungr et al. 1984), and
an equation relating the super-elevation of deposits on bends (Johnson
and Rodine 1984).

The velocity-head equation is given by:

$$h = v^2/2\,g \qquad\qquad (eq.\ 5.1)$$

where h is the run-up height (m), v is velocity ($m\,s^{-1}$) and g is gravitational
acceleration (9.807). Rearranging for v, gives;

$$v = (2\,gh)^{0.5} \qquad\qquad (eq.\ 5.2)$$

Velocity along the beck can be estimated using field evidence of the super-
elevation of debris on the bends of the channel. Where the channel slope
is less than 15° this can be calculated from the equation:

$$v = [g'\Psi'\tan\beta]^{0.5} \qquad\qquad (eq.\ 5.3)$$

where Ψ is the radius of curvature of the centre-line of the channel and
β is the cross channel tilt of the preserved deposits (Johnson and
Rodine 1984).

Results based on these equations suggest at the point of entry into the
valley the unconstrained run-up velocity was $10.4\,m\,s^{-1}$. With distance
downstream the velocity rapidly declines; 60 metres downstream of
the valley-entry point the velocity is reduced to $6.5\,m\,s^{-1}$ and at 200
metres downstream it falls to $4.1\,m\,s^{-1}$. Assuming this trend to continue,
by 400 metres velocities are likely to be low and transport will be

dominantly fluvial as the slurry is rapidly dispersed by the river flow. Several assumptions are explicit in these calculations. Firstly, the velocity head approach neglects friction which tends to result in an overestimate of the run-up speed (Hungr et al. 1984). Secondly, velocity determinations using this approach assume the peat flow is moving as a debris-type flow. Although the rheological properties of peat slurry are generally poorly defined (Luukkainen 1992) several authors have argued that these flows can reasonably be termed debris flows (Bishopp and Mitchell 1946; Colhoun 1965; Carling 1986a). Eyewitness accounts describe the in-channel flows as 'a wall of mud with the appearance and consistency of chocolate sauce' (Carling 1986b; Archer 1992: 184), and as such have characteristics more akin to hyperconcentrated flows. Morphological evidence including lateral ridges (levees), over-passing of intact vegetation, super-elevation on the valley sidewalls and overbank, lobate deposits from multiple peat slides and bog flows strongly supports a debris-flow movement mechanism.

Observations of the failed peat mass, the form of the peat blocks and smaller-scale sedimentary features in the runout track are often very useful in providing evidence that can be used to infer the dynamics of failure. For example, Warburton et al. (2003) carefully mapped peat block characteristics in the Harthope peat slide in the North Pennines and were able to determine the general nature of the flow from the size distribution of peat blocks, small-scale flow-direction sedimentary indicators and the tilt and orientation of peat blocks.

5.3.4 Secondary tension and compression features

In association with peat mass movements there are often a variety of features related to tensile and compressive stresses within the disturbed peat mass. These usually consist of: (1) tears – surface displacements of peat resulting in the rupture of the surface vegetation and exposure of underlying peat (Figure 5.5d) and/or (2) cracks – vertical or near vertical structures dissecting the peat down to the underlying substrate (Figure 5.5e). These two features are not mutually exclusive and peat tears may develop into fully developed cracks.

Tension cracks may form single features or may occur in groups which often show several distinctive general geometries such as ranks of parallel cracks or concentric crack systems around the head scarp of a peat failure. Differences between crack systems generally reflect the degree of vertical displacement and rotation of the cracked peat. Small-scale graben-like features may develop with backwards rotation of the peat block. Alternatively cracks that form patterns such as crow's feet or spider cracking are different inasmuch as they show loci of crack development, for example displacement about a resistant mound in the substrate. The initiation of cracking and its propagation is a precursor to peat raft and block development within the failed peat mass. The size of peat blocks and geometry of large peat rafts is related to the spacing of crack networks. It appears that crack geometries are both related to the material and geotechnical properties of the peat and the local depth of the peat deposit.

Compression ridges, consisting of folded anticlinal structures in the peat mass (Figure 5.5f) are often found immediately adjacent to the site of mass movement. These tend to occur where compressive stresses in the failed mass develop along the lateral margins of the failure zone or at the distal end of the failure. These are particularly well-developed where there is a distinct two-component peat stratigraphy with a fibrous peat overlying a more amorphous sub-layer. These pressure ridges are fairly common and may form within the zone of disturbance but also occur both as leading edge (transverse) ridges and marginal longitudinal ridges. Compression ridges can run for several tens of metres and in extreme cases several hundreds of metres across slopes. Heights vary from a few decimetres to in excess of 2 metres (Figure 5.5f). Larger amplitude ridges are generally fractured along their crests as over thrusting occurs. Where the peat is sufficiently soft (e.g. runout into boggy ground or mires), multiple parallel transverse ridges may occur often showing a decline in amplitude with distance downslope as the momentum transfer from the peat slide is damped out. Lateral compression features are less common but are generally associated with horizontal displacement of the failed mass causing marginal over-thrusting and rippling. In some failures there is evidence that this occurs during the early phase of failure prior to lateral shearing as the main failure is excavated of material.

These small-scale features are important for a number of reasons:

1   they are useful indicators of the stress history of the peat mass;
2   the cracks break-up the hydrological continuity of the peat mass and allow new surface and subsurface drainage systems to develop;
3   they provide new micro-habitat within the peatland system;
4   they present a hazard to livestock.

One interesting question that remains to be answered: is a cracked peat mass more stable than an intact one? There are a number of apparently contradictory factors to consider. Firstly, given that most peat failures are hydrologically controlled, cracking of the peat mass breaks up the hydrological continuity of the slope and encourages slope drainage and dissipation of pore pressures. Secondly, a limited degree of desiccation cracking may actually promote delivery of surface water to the subsurface hydrological system promoting elevated pore pressures. This is something which has been hypothesized for a number of peat slides (Alexander et al. 1986; Hendrick 1990), and is support by the common association of peat mass movements with the first large storm following a drought period (Wilson and Hegarty 1993; Mills 2002). Thirdly, around a peat fracture mass erosion rates may be locally enhanced due to secondary processes, for example small soil collapses and surface weathering. Therefore it is probably the nature and extent of cracking that governs the overall stability of the peat mass.

5.3.5   Bog burst and peat slides – are they different?

It is clear from the preceding discussion that several distinctive morphologies of peat mass movement are preserved post-failure. If we examine the basic characteristics (slope angle and volume of material displaced) of peat mass movements in the British Isles and plot these with respect to altitude (Figure 5.6), it is possible to see some general characteristics emerging (Mills 2002). Firstly, the distribution of bog bursts tends to occur at lower slope angles than peat slides. Minimum slope angles may be as low as 1–2° whilst at peat slides minimum angles are 3–4°. The majority of peat slides occur on slopes of between 5–20° compared to bog bursts whose range is more typically 2–8° (Figure 5.6a). In terms of volume of failure, bog bursts are generally larger than peat slides (Figure 5.6b). Peat-slide volumes tend to range from 10^2 to 10^5 m^3 but bog bursts are usually greater than 10^3 and can be as large as 10^6 m^3. The large volume of bog bursts in comparison to peat slides is primarily a virtue of the deeper peat found at these sites. Figure 5.7 shows this graphically; slide sites are very strongly clustered around peat depths of 1 to 2 metres but bursts occur in deeper peat (1.5 to 6 m) and in some cases very deep peat of over 9 metres. These simple comparisons clearly suggest that there is a link between peat depth and type of peat mass movement. Thus a more detailed investigation of the peat stratigraphy may reveal a link between these variables and material properties.

Figure 5.8 shows a summary of the peat stratigraphy reported for bog bursts (British Isles) and field surveys of peat-slide scar stratigraphy in the North Pennines (Mills 2002). Burst stratigraphies show a great range and maximum depth of peat compared to the peat-slide sites. Peat slides vary in depths between 0.5 to 1.5 metres whilst bog bursts are as deep as 3 metres (also see Figure 5.7). The characteristic stratigraphy of the slide and burst stratigraphic logs is markedly different. Bog bursts have a 'two-

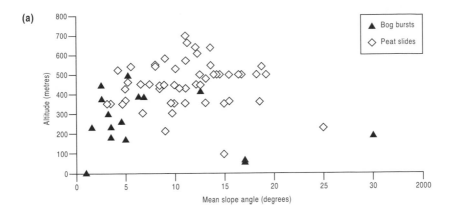

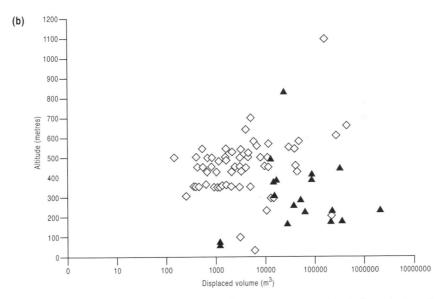

Figure 5.6   Variation in peat mass movement attributes with respect to altitude. Examples from the British Isles (n = 133). (a) Mean slope angle for bursts and slides. (b) Displaced volume of peat for bursts and slides (Mills 2002)

layer' (two-phase) stratigraphy which generally consists of less humified fibrous peat (up to 1 m in thickness), overlying an equivalent or deeper layer of amorphous well-humified peat. In slide stratigraphies fibrous, less humified peats dominate throughout the peat profile. This evidence clearly demonstrates an important link between the material properties of the peat mass and the form and mode of failure.

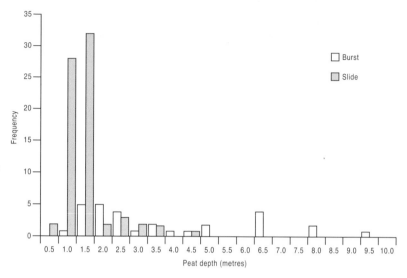

Figure 5.7 Maximum depths of peat recorded at peat mass movement sites. Data are for British Isles. Divided between burst and slide sites (modified from Mills 2002)

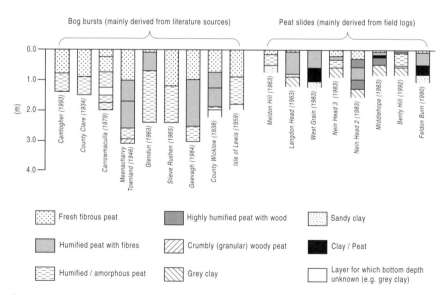

Figure 5.8 Summary of general peat stratigraphy found at British burst and slide peat mass movement sites. Burst stratigraphies are largely based on literature accounts, slide sites are based on field observations (modified from Mills 2002)

5.4   Mechanism of Peat Failure

The fundamental requirement for the stability of any slope is that the shear strength of the hillslope material must be greater than the applied shear stress (Graham 1984). Failure of a hillslope will result if there is a significant decrease in the shear strength of the soil or an increase in the shear stresses acting on the slope (Duncan and Wright 2005). For peat slope failures, several mechanisms have been proposed. These include: shear failure by loading, buoyancy effects, basal liquefaction and surface or marginal rupturing (Mills 2002; Warburton et al. 2004).

Sites of peat mass movement share several characteristics which predispose them to failure (Tomlinson and Gardiner 1982; Carling 1986a). These preconditions all relate to hillslope hydrological processes either directly or indirectly (Warburton et al. 2004) and are:

1   impeded drainage caused by a peat layer overlying an impervious clay or mineral base (hydrological discontinuity);
2   a convex slope or a slope with a break of slope at its head (concentration of subsurface flow);
3   proximity to local drainage either from flushes, pipes or streams; and
4   connectivity between surface drainage and the peat/impervious interface.

Table 5.2 (modified from Warburton et al. 2004) summarizes the main failure mechanisms described in accounts of peat slides and bog bursts; examples of each type of event are given. For all types of peat mass movement an increase in moisture content near to the peat/substrate interface appears to be the major control upon failure. In the case of shear failure, water may act to increase the shear stresses on the peat (or substrate) in three ways (Warburton et al. 2004). Firstly, by loading the basal material through saturation of the overlying peat exceeding the frictional resistance of the soil and causing it to fail (Landva and Pheeney 1980). This process would be most likely to occur in the peat substrate (Carling 1986a), as the concept of tensional rather than frictional resistance is favoured for the cellular structure of peat (Helenelund and Hartikainen 1972; Wilson 1972). Secondly, shear stresses may be generated by the presence of hydrostatic loads acting horizontally from infilled cracks, ponds and artificial drainage lines (Bowes 1960; Wilson and Hegarty 1993). Finally, some early reports of peat mass movements rather speculatively suggested a hydraulic mining effect caused by the force of rain upon the bog surface (Muschamp Perry 1893; Hudleston 1930).

Table 5.2 Examples of failure mechanisms proposed for peat mass movements (table modified from Warburton et al. 2004)

Failure mechanism	Description	Example
Shear failure by loading	• Increase in overburden – weight of absorbed water (rainfall, snowmelt) or snow	Slieve-Rushen (Colhoun 1965)
	• Increase in shear stress – hydrostatic pressure generated by water-filled cracks, ponds and lochs	Isle of Lewis (Bowes 1960)
	• Catastrophic loading	Dow Crag (Hudleston 1930)
Buoyancy effect	• Generation of artesian pressures	Danby-in-Cleveland (Hemingway and Sledge 1945)
	• Increase in interstitial pore-water pressure and reduction in cohesion	Slieve-an-Orra (Tomlinson and Gardner 1982)
Liquefaction	• Basal peat slurried by increased water content (exceed liquid limit)	Green Gorge (Selkirk 1996)
	• Basal clay slurried by organic acid dispersal (passing of liquid limit)	
	• General increase in basal moisture content by routing of artificial drainage	Carntogher (Wilson et al. 1996)
Surface rupture	• Swelling of basal peat leading to rupture of the drier surface	Glendun (Colhoun et al. 1965)
	• Contraction of surface peat during drought	Glencullin (Delap et al. 1932)
	• Long-term depth creep inducing surface rupture or shear failure	Moanbane (Mitchell 1938)
Margin rupture	• Removal of underlying support by stream action – release of basal peat	Powerscourt (Delap and Mitchell 1939)
	• Removal of underlying support by peat cutting	Knocknageeha (Praeger 1897a)

5.4.1   Speed of failure and movement

Peat mass movements have frequently been described as flows, slurried masses or 'liquid in nature' (Bailey 1879; Hemingway and Sledge 1945). Alternatively descriptions have also referred to en-masse movements (Beven et al. 1978). However there remain few direct accounts of movement (Kinahan 1897a). Observations of the speed of failure are valuable because they provide important clues to the nature of the failure mechanism. Rapid failures are generally associated with a sliding translational failure, whilst slower rates may be more closely related to internal deformation of the peat body. Direct observations of peat mass movements are rarely documented, therefore the evidence is fairly limited; however, in some cases evidence may be reconstructed from the morphology of the slide deposits (Warburton et al. 2004). Sollas et al. (1897: 498–9), in their review of bog flows in Ireland, were the first to generalize on the relationship between the velocity of the mass movements and the rheology of the flow. They comment: 'The rate of flow is evidently a function of the slope of the ground and viscosity of the fluid, and the latter depends on the ratio between the amount of water and solid contents present in the moving material.' They go on to comment on the significance of the upper peat stratigraphy in controlling the mode of failure. Where a large proportion of solid material is present a translational slide (landslip) is likely but where a large liquid content is present a rapid flow occurs (Sollas et al. 1897: 499).

Available evidence for the speed of peat mass movements is summarized in Table 5.3. The main conclusion that can be drawn from this compendium is that the estimated rates of movement are highly variable. Much of the variability relates to the definition of what is meant by 'the period of the failure' or 'area of the flow' which is being described. For example extremely slow rates are often associated with pre-failure creep (Sollas et al. 1897) or post failure readjustment (Praeger 1897b); moderate rates are usually directly observed in the zone of failure (Warburton 2002); and rapid rates in the runout track (Alexander et al. 1986). Although the data are limited, bursts appear to fail over longer timescales than peat slides.

5.5   Significance of Surface Hydrology in Peat Failures

Warburton et al. (2004) have demonstrated that hydrological processes are fundamental in determining the spatial and temporal occurrence of peat mass movements.

Table 5.3 Summary of evidence for speed of peat mass movement failures

Failure	Failure type	Observed speed of failure*	Source
Tutoh River, Sarawak, Malaysia, 1961	Bog flow	Duration 60 minutes	Wilford 1965
Bog of Rine, Camlin River, Co. Longford, 1809	Bog flow	Duration several hours	Sollas et al. 1897
Ballykillon, Kilnalady, King's County, 1821	Bog flow	$1.8\,m\,hr^{-1}$ Crept for a month	Sollas et al. 1897
Stanley, Falkland Islands, 1878	Bog flow (liquid peat)	Travelling downhill at 6.4 to $8\,km\,hr^{-1}$	Bailey 1879
Hart Hope, North Pennines, 1995	Peat slide	Velocity in flow track $10\,m\,s^{-1}$	Warburton et al. 2003
Glen Culliun, Co. Mayo, 1931	Bog flow	Duration of flow approximately 7 hours	Delap et al. 1932
Dunmore, Co. Galway, 1873	Bog flow	Slowly moving mass continuing for 11 days	Praegar 1897
Unspecified, Ireland, 1986	Bog flow	Duration 2 days, fluctuating velocity and moving at a rate faster than walking speed	Tallis 2001
Noon Hill, North Pennines	Peat slide	$1\,m\,hr^{-1}$	Tallis 2001
Straduff, Co. Sligo, 1984	Bog flow	Average velocity in the runout close to the source varied between 3.6 to $6.2\,m\,s^{-1}$	Alexander et al. 1986 Coxon 1986
Carlcotes, near Huddersfield	Peat slide	Failure occurred at walking speed	Warburton 2002 pers. comm.

* Values converted to metric units from original data.

5.5.1 Water content, pore pressures and volume changes

At the scale of the soil profile the special hydrological properties of peat, in particular shallow water-tables and low soil hydraulic conductivity, offer important clues to failure mechanisms. Increases in basal moisture content may lead to buoyancy effects or liquefaction (by exceedence of the liquid limit) of the basal material (clay or peat) to produce a zone of failure. Pore-water pressures are generally low and vary little in the peat profile and throughout the year (Warburton et al. 2004). It is clear from observations of subsurface flow that significant volumes of runoff reach considerable depths in the soil (Germann 1990; Holden and Burt 2002a). Carling (1986a) has described the build up of artesian pressures in the lower slopes of failures in the Pennines. Elsewhere, Gilman and Newson (1980) note high pressures in pipe networks in peat catchments in Wales. Macropores are important in delivering surface runoff to deeper parts of the profile and in some instances high water pressures may develop in natural soil pipes.

On a micro-scale, the disruption of cell structure by loading, creep or seepage pressures (Mitchell 1938; Wilson 1972; Crozier 1986) may liberate water held within plant cells into void water, thus increasing pore-water pressures and buoyancy effects (Glynn et al. 1968). Such behaviour may lead to the phenomenon of 'quaking' or 'floating' peat, where the upper peat mass is supported by a subsurface water body (Price and Schlotzhauer 1999). Alternatively 'reversal of phase' (solid–liquid) associated with disturbance/remoulding has been suggested as a cause of failure (Bishopp and Mitchell 1946). The period over which disturbance takes place may range from the very short term once a failure has initiated (Alexander et al. 1986), to a much longer period of remoulding via basal creep (Crozier 1986; Table 5.3). The actual change in moisture content required may be negligible. Hobbs (1986) notes liquidity ratios of peat very close to those of Norwegian quick clays, whilst Delap et al. (1932) has described failed material as 'water with peat in suspension'. Hanrahan (1954) found water content and material strength were directly related, suggesting that bog peats have a negligible compressive strength because they are so close in character to a liquid. The susceptibility of peat to swell under increasing moisture content and contract on drying (Ingram 1983; Price and Schlotzhauer 1999) may be partly responsible for surface rupturing during bog bursts. Colhoun et al. (1965) described a bog burst in Glendun, Ireland, where basal swelling ruptured the drier peat surface, while Delap et al. (1932) attributed a similar event to the drying of surface peat, causing tension between it and the wetter, lower peat. Repeated cycles of desiccation, causing cracking, may allow tensile stresses to produce permanent planes of weakness in the peat mass (Bower 1959).

5.5.2 Rainfall

A common feature of most failures is the presence of high rainfall prior to the event (Beven et al. 1978; Caine 1980; Rogers and Selby 1980). Seasonal analysis of recorded UK and Irish peat mass movements shows that roughly half occur in the late summer months of July and August, with a few scattered through the winter months of November, December and January (Warburton et al. 2003). This distribution predominantly relates to summer convective storm activity and the effects of prolonged autumn and winter rainfall. Peak daily rainfall amounts recorded at the time of peat failures range between as little as 2.45 millimetres and as much as 203 millimetres. Monthly rainfall totals usually reveal greater than average percentages for the month of failure (Colhoun et al. 1965; Alexander et al. 1985), but interestingly in some cases, considerably below average values for the months prior to failure (Werritty and Ingram 1985; Alexander et al. 1986; Hendrick 1990; Wilson and Hegarty 1993). This may indicate increased potential for crack formation by desiccation and for the build-up of stresses between upper and lower peat layers. It is not always the case that high intensity rainfall precedes failures, and in some cases, the absence of rainfall has been noted (Mitchell 1938).

5.5.3 Slope drainage

The association of peat mass movements with natural lines of drainage has been made on a number of occasions. Selkirk (1996) found that five out of seven peat slides on Macquarie Island took place along vegetated drainage lines. Mills (2002) examined records of slope drainage for bog burst and peat slides reported in the literature and found a large proportion of the peat-slide failure sites had drainage direct into the heads of the failure (Table 5.4). It is therefore concluded that local drainage is a highly significant factor in most localities. Often drainage lines within peatlands form very wet, frequently ponded peat-flush zones. These flush zones typically have very small vertical hydraulic gradients with flow nets occasionally indicating an upward flux from within the peat mass toward the surface, probably as a result of the combination of a head of water from upslope and a fairly impermeable peat layer.

Table 5.4   The importance of slope drainage in the failure of peat hillslopes (modified from Mills 2002). Note all values are % surveyed from the literature on peat slide failures (n = 133). Values exceed 100% because several drainage features are noted at a single site

	Drainage conditions at the failure site					
	Into head	*Below site*	*Into margins*	*Within failure zone*	*Drainage unreported*	*Not known*
Slides	3.8	6.3	3.8	30.4	46.8	8.9
Bursts	2.0	12.2	2.0	10.2	26.5	46.9

Crozier (1986) cites such seepage pressures as a means by which changes in moisture content may act to destabilize a slope. At most peat failure sites, point and diffuse water seepage is present in both the peat and substrate. Subsurface soil pipes can be found at the contact between different layers of peat, at the peat substrate interface and within the substrate itself. Similarly, more diffuse seepage occurs along soil unit boundaries. Evaluating the hydrological significance of these discontinuities within the peat and the significance of the peat-substrate interface in failure is difficult given the natural variability in stratigraphy and material properties.

5.6   Stability Analysis and Modelling of Peat Mass Movements

Most reports of peat mass movements use evidence based on post failure landforms (Mitchell 1935), eye-witness accounts (Kinahan 1897b), or local rainfall records (Hudleston 1930; Colhoun et al. 1965) to determine the cause and mechanism of failure. Analysis has largely been superficial. This reflects a lack of rigorous geomorphological and engineering approaches to the problem and the difficulties of adequately describing the material properties of peat. Within the engineering literature peat has been noted as a particularly complex engineering material (Barden and Perry 1968). A few studies have included a more rigorous geotechnical approach; these include Ward (1948) and Hendrick (1990), who performed stability analyses on unstable peat and Carling (1986a), Dykes and Kirk (2000, 2001), and Warburton et al. (2003) who derived factors of safety for the failed clay beneath peat slides. Dykes and Kirk (2001) used a simple finite element model to examine the hydrological conditions at a small peat-slide site in Northern Ireland.

Stability analysis has therefore not been routinely undertaken for peat mass movement failures. Estimating the shear strength of peat is a problem which engineers have struggled with for several decades (Landva and Pheeney 1980). If peat is treated as a cohesive material, shear strength can be expressed in terms of total stress of the material as:

$$s = c + \sigma\tan\phi \qquad \text{(eq. 5.4)}$$

where s is the shear strength, c is cohesion, σ is the total stress and ϕ is the angle of internal friction.

Estimating c and ϕ for peat soils is notoriously difficult and application of this geotechnical modelling approach is hindered by two factors. Firstly, published values of the shear strength properties of peat (cohesion and angle of internal friction) are relatively few; and secondly, testing of peat using standard geotechnical tests is fraught with problems especially when trying to remove intact samples from the field without disturbance or field testing using conventional vane and penetrometer tests in the presence of multiple fibres (Jennings 2006). Published data (Kirk 2001; Mills 2002; Dykes and Kirk 2005) suggest values for cohesion in the range of 0 to 18 KPa and internal friction angles of 5 to 48°. These values vary with the vegetational composition of the peat (particularly fibre content) and the degree of humification. Compared to other slope materials the inherent variability in these basic material strength parameters makes the material particularly difficult to characterize and hence, adequately parameterize in geotechnical models. The strength properties of peat, like many other types of soil, are very complex and subject to changes over time through a variety of factors including consolidation, wetting/drying, weathering and creep. Some of these characteristics in very humified peat can approximate the behaviour of clays but in fibrous peat, strength properties are considerably more difficult to characterize.

In cases where an engineering stability analysis has been applied the infinite slope model has been used (Carling 1986a; Dykes and Kirk 2001; Warburton et al. 2003). The infinite slope model assumes failure occurs parallel to the slope as a planar translational slide where the length of slope is large compared to the depth of failure. This is appropriate for many shallow peat mass movements. The stability of the peat slope is usually assessed by calculating a Factor of Safety (F), which is the ratio between the sum of resisting forces (shear strength) and the sum of the instability forces (shear stress):

$$F = \frac{c' + (\gamma - h\gamma_w)z\cos^2\beta\tan\phi'}{\gamma z\sin\beta\cos\beta} \qquad \text{(eq. 5.5)}$$

where c' is the effective cohesion, γ is the unit weight of saturated peat, γ_w is the unit weight of water, h is the height of the water-table as a proportion of the peat depth, z is the peat depth in the direction of the normal stress, β is the slope angle from the horizontal and φ' is the effective angle of internal friction.

Results from these analyses generally indicate that factors of safety are too high for failure to occur unless some mechanism for weakening the slope materials is invoked (Carling 1986a; Warburton et al. 2003). Dykes and Kirk (2001 and 2006) used a more sophisticated modelling approach involving the 'method of slices' whereby the slope is partitioned into a set of representative segments and the infinite slope model is applied to each. This has the benefit of indicating where on a slope failure is most likely, but is still subject to the other limitations previously mentioned. Using this method Dykes and Kirk (2006) found that the slice with the lowest factor of safety generally corresponded with the point of failure as defined by the field evidence. Importantly failure at the peat-substrate interface has rarely been modelled despite the wealth of field information suggesting this occurs (Mills 2002).

5.7   The Changing Frequency of Peat Mass Movements Over Time

Pearsall (1950) was one of the first authors to recognize the general importance of peat mass movements for landscape development. He suggested that peat mass movements were becoming less common than previously reported because of the widespread drainage of peatlands and changes in land-use which have resulted in vegetation that is more resistant to lateral tearing. Pearsall bases this assumption on the anecdotal evidence that many of the larger peat bogs are often referred to as flows. This contradicts recent evidence of a apparent increased frequency of peat mass movements in recent decades (Mills 2002; Figure 5.9).

Figure 5.9 illustrates an increasing frequency of landslides recorded in British peatland environments and the associated frequency of triggering events, such as intense rainstorms. Large upland storms are particularly important because one severe event may trigger a single failure in a particular locality, or trigger multiple failures. The failures that populate the curve represent some 80 per cent of the total global record of landslides in peat (Mills 2002). The pattern in Figure 5.9 appears to indicate that the rapid increase in landslide frequency in recent years is associated with a growing occurrence of large failure clusters during individual severe storm events, rather than an increasing number of single dispersed failures. This indicates that landslide events in peat are becoming more

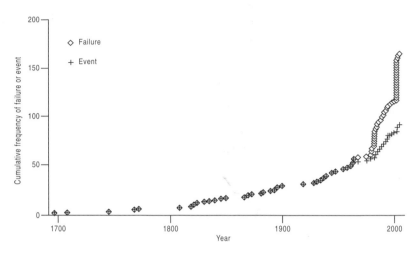

Figure 5.9   Changing frequency of peat mass movements in British peatland environments. Events are defined as the storm which triggered a particular slope failure or group of failures (modified from Mills 2002). A failure is defined as an individual mass movement site

extreme, resulting in greater societal and economic loss per event. A similar increase in the frequency of global landslides in peatlands resulting from environmental change (possibly driven by permafrost degradation) would have a significant role in degrading the global peat resource. Recent work in the Arctic suggests this may have already begun with the frequency of active-layer detachment slides increasing in recent decades (Lewkowicz 1992). Comparison of this peat landslide record with other landslide datasets would be a useful means of determining whether similar patterns exist in non-peatland environments. However, historical records must be interpreted with some caution because of uncertainties related to the reporting of landslide occurrence with early storm events in the series (McEwen and Withers 1989). Palaeoecological evidence suggests variations in peat mass movement activity have operated over longer timescales than indicated in Figure 5.9 (Tallis et al. 1987; Ashmore et al. 2000; Caseldine and Gearey 2005); hence recent data need to be placed in the context of a longer time series.

Mills (2002) attempted to calculate the significance of peat slides in the overall sediment budget of the North Pennines peatland over the last century by comparing sediment losses from peat slides with background fluvial activity. Calculations showed that just 3 per cent of the total sediment yield could be associated with rapid peat mass movements over the last 100 years. Although this in a regional context is relatively minor, locally peat-slide events remain highly significant and may have disastrous short-term impacts on fluvial systems (McCahon et al. 1987).

5.8   Summary and Overall Framework

This chapter has demonstrated that rapid peat mass movements can be divided into slides and bursts defined by their main mode of failure and initial movement. Creep is also significant in the wider peatland landscape and may be important in weakening slopes. Available evidence suggests creep rates in peat tend to be higher than in mineral soils. Material properties, local topography and hillslope hydrological controls exert a strong influence over the nature of failure and the spatial distribution of these mass movements. Failure mechanisms are not clearly established. Several competing hypotheses exist to describe mechanism of failure (Table 5.2). These include shear failure by loading, buoyancy effects, basal liquefaction or surface and marginal rupturing. All mechanisms are intimately linked to hillslope hydrology and the special physical properties of peat.

Defining the location of the shear plane in the hillslope of a peat mass movement is often the key to understanding the mechanism of failure at a particular site. Based on available evidence three modes of failure can be generalized. The three modes include: failure at the peat/substrate interface (Figure 5.10a); failure within the underlying substrate (Figure 5.10b); and failure within the peat mass (Figure 5.10c). These simple scenarios do not describe all examples of peat failures but rather provide useful conceptual models which could be tested with generic modelling approaches (cf. Dykes and Kirk 2001).

In conclusion, Figure 5.11 shows a simple framework which can be applied to the classification of peat mass movements. This chapter has demonstrated some of the distinct characteristics of the morphology of peat mass movements. These might be characterized as bog bursts and peat slides which appear to be two end members of a range of forms of peat mass movements. It is also evident that peat mass movement runout features are hard to differentiate from initial source conditions because rapid degeneration of peat in transport leads to similarity of deposition forms in the runout track. Given the inherent complexities of peat and the enormous variety of hydrological settings where peat failures occur it is recommended that when classifying peat mass movements a simple scheme like the one outlined in Figure 5.11 is adopted. In closing it should also be recognized that peat mass movements provide major contributions to the overall peatland sediment system, particularly over short timescales and in a local context, and have been cited as important mechanisms for

(a)

(b)

(c)

Figure 5.10 Examples of particular failure modes in typical peat slope settings. (a) Peat slide failure at the peat/substrate interface, Harthope, North Pennines, UK, 1995. (b) Peat slide failure within the substrate below the peat, Dooncarton, Co. Mayo, Western Ireland, 2003 (photograph: Alan Dykes). (c) Bog burst failure within the peat, Geevagh, Co. Sligo, Ireland 1984

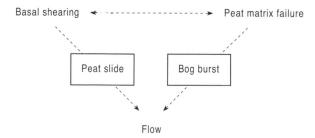

Figure 5.11   A simple framework for the definition of peat mass movements. Peat slides and bog bursts are seen as two distinctive morphologies arranged as the end members along a continuum of forms. Once movement is initiated, rapid degeneration of the peat mass during flow results in a convergence of form in the runout features

gully initiation. However, over the longer decadal timescale peat mass movements represent a small fraction of sediment transfer compared to fluvial processes operating in peat catchments, and rapid re-vegation and stabilization of eroded scar surface rapidly reduces short-term sediment delivery.

Chapter Six

Wind Erosion Processes

6.1 Introduction

Wind erosion is the process by which loose surface material is picked up and transported by the wind (Wilson and Cooke 1980). Peat is a geological material which has a low bulk density. Typically saturated peat has a bulk density close to $1 \, mg \, m^{-3}$ but field bulk densities will vary greatly with dry bulk densities an order of magnitude less than the saturated state (Galvin 1976; Nichol and Farmer 1998; Campbell et al. 2002; Table 6.1). These characteristics mean peat is potentially very susceptible to wind erosion and aeolian transport. Furthermore, given the propensity for upland peatlands to occur in exposed and often open settings, the potential for entrainment of surface material is large.

Over 7.5 per cent of the British Isles $(22,500 \, km^2)$ is covered in blanket peat (Tallis 1998), and wind erosion has been widely reported in many upland peatlands of the UK (Chapter 1). Given the combination of high winds and exposed upland summits it is therefore not surprising that many authors have argued for the importance of aeolian processes in the erosion of peat, often in association with other disturbance factors such as fire or over-grazing (Radley 1962; Kinako and Gimingham 1980; Anderson and Radford 1994). Until recently, upland peat erosion by wind had never been directly measured in the UK despite its importance being recognized for several decades (Rastall and Smith 1906; Samuelsson 1910; Bower 1960a; Radley 1962, 1965; Barnes 1963). Past research had been qualitative in its approach and there are still very few direct quantitative measurements of rates of wind erosion or the dynamics of the process (Campbell et al. 2002; Warburton 2003; Foulds and Warburton 2007a and b). The aim of this chapter is to summarize existing knowledge of upland peat wind erosion. The main objectives are: to review past

Table 6.1   Estimates of the wet and dry bulk densities of peat and mineral soils

Peat type	Wet/field bulk density (Mgm⁻³)	Dry bulk density (Mgm⁻³)	Source
Milled peat (von Post* 3–6)	0.19–0.24	0.04–0.08	Campbell et al. 2002
Loose peat (von Post 3–6)	0.23–0.40	0.1–0.23	Campbell et al. 2002
Crusted peat (von Post 3–6)	0.21–0.36	0.1–0.22	Campbell et al. 2002
Irish blanket peat	1.02	0.07	Galvin 1976
Welsh bog peat	0.99–1.16	0.09–0.16	Nichol and Farmer 1998
Peat soils (von Post 1–10)		0.05–0.2	Egglesmann et al. 1993
Mineral soil (A horizon)		1.0	Egglesmann et al. 1993
Mineral soil (B horizon)		1.5	Egglesmann et al. 1993

* von Post (1924). Humification scale for peat. Values range from 1 (no decomposition with clearly visible plant remains) to 10 (complete decomposition with no discernible plant structures).

research on wind erosion in upland environments; describe the basic processes which operate; examine how wind erosion is influenced by dry conditions (drought); and discuss the quantitative measurement of wind erosion and peatland aeolian process dynamics.

6.2   The General Significance of Wind Erosion in Upland Peatlands

From the perspective of erosion of agricultural land the most severe cases of wind erosion occur in lowland England and Wales, but significant problems can also occur in upland areas where over-grazing or recreational activities have removed the vegetation cover (Ministry of Agriculture Fisheries and Food 1993). The loss of resource through erosion has widespread consequences for upland agriculture and recreation. Upland peat moorlands are also central to hill-farming economies thus erosion of peat can be a potentially serious problem for land managers where it is actively occurring, requiring knowledge of the processes responsible (Ellis and Tallis 2001).

In a UK context Ballantyne and Harris (1994: 23) state: 'Perhaps the most notable feature of the upland climate of Great Britain is the strength and persistence of the wind, which reflects concentration and acceleration of airflow as it passes over mountain barriers.' Similarly, aeolian landforms on the British uplands have been widely recognized for several decades but have been primarily associated with cohesionless sand-rich regoliths (Ball and Goodier 1974; Goodier and Ball 1975; Birse 1980; Pye and Pain 1983; Ballantyne and Whittington 1987). These include deflation surfaces, wind-patterned ground and turf-banked terraces.

In terms of UK peat, one of the earliest references to the significance of wind erosion was Samuelsson (1910) who observed that denudation of Scottish peat mosses was dominantly by wind erosion. It has been generally found that on lower angled slopes of less than 5°, stream erosion leads to the development of closely-spaced reticulate islands which give way to linear gully forms as hillslope gradients increase (Bower 1960b). However, in summit locations, distinctive erosion surfaces known as 'peat flats' have been reported (Bower 1960b; Radley 1962; Tallis 1965). These are isolated from the stream drainage, and wind-driven rain and wind action are believed to be the dominant processes of erosion (Radley 1962). However, in the 1960s there was a major disagreement between geomorphologists in the Britain about the significance of wind in eroding upland peatlands (Barnes 1963). Bower (1961a and b) viewed fluvial processes as the significant driving force behind upland erosion. However, Radley (1962) disagreed and saw aeolian processes as the most significant agent of summit degradation (Figure 6.1b): 'Wind action will be frequent and vigorous on high, bare, exposed Pennine summit areas' (Radley 1962: 48). Part of the reason for these diverse opinions stemmed from the experience of the two observers. Radley was strongly conditioned by observations of erosion following moorland fires in the North York Moors, where wind ravaged the exposed and friable fire crust and produced extensive aeolian deposits including small rippled dunes (Radley 1965). Bower's work was more concerned with gully erosion and hence favoured fluvial processes (Figure 6.1a and d).

Direct measurements of wind erosion rates in the UK are very rare and only usually occur in areas of intensive agriculture where wind erosion is a known problem (Morgan 1980). Estimates for individual storms are often quoted but annual rates of soil loss per unit area are extremely sparse. Pollard and Miller (1968) describe wind erosion of peat soils in the East Anglian Fens in 1968. Arden-Clarke and Evans (1993) and Boardman and Evans (1994), both quoting from Wilson and Cooke (1980), provide maximum annual sediment yield estimates of between 20 and 44 t ha^{-1} for areas affected in the Vale of York, Nottinghamshire and North Norfolk. More recently Chappell and Warren (2003) use ^{137}Cs

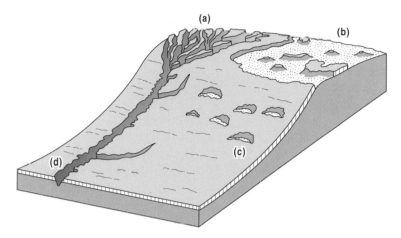

Figure 6.1   Schematic diagram of peat erosion from Radley (1962). The diagram shows two forms of summit erosion produced by severe dissection (a) and wind erosion (b). Arcuate tears, indicative of peat creep are shown on the hillside (c) and linear gully dissection on the slopes (d)

mapping to estimate wind erosion soil loss in East Anglia. Based on this technique, net annual soil erosion is estimated at between 2 to 5 t ha^{-1} on a light sandy soil on farmland. Using sediment-yield data from Moss Flats in the North Pennines for 1999 and 2000, Warburton (2003) calculated annual net erosion rates are 0.46 and 0.48 t ha^{-1}, respectively. Foulds (2004) working at the same site in 2004 calculated an annual erosion rate of approximately 0.24 t ha^{-1}, approximately half the yield of the earlier estimates. Therefore erosion from upland peat soils is generally an order of magnitude less than the erosion rates reported above and in extreme cases two orders of magnitude lower. However these rates should be viewed carefully because the dry bulk density of peat is also an order of magnitude less than most mineral soils (Table 6.1). By mass these figures correspond to an average annual surface lowering of approximately 0.5 millimetres which is equivalent to a surface lowering of approximately 3 millimetres in the field. This means that in terms of the volume of soil loss, values are much closer to the soil loss experienced with lowland agricultural soils than first would appear.

Furthermore, comparisons of the significance of wind erosion of upland peat with other similar studies cannot be readily undertaken because the results reported are often based on different measurement methods. Selkirk and Saffigna (1999) discuss wind and water erosion of peat on sub-Antarctic Macquarie Island but their approach using erosion pins provides an estimate of surface lowering but not an estimate of the relative importance of wind and rain or the horizontal mass sediment flux.

In an international context there have been very few studies of the wind erosion of peat. In some cases the erosion of peat by wind has been termed 'turf exfoliation' (Troll 1944; Seppälä 2004). Exposed peat faces are susceptible to erosion by mineral particles and wind-blown ice crystals (Åhman 1976; Seppälä 2001). Eventually abrasion of the free soil faces and stripping of surface material by wind results in the collapse and loss of the overlying turf cover. Rates of soil loss can be extremely high. Seppälä (2001) notes the removal of 0.4 metres of peat by winter winds in Finland and Cummings and Pollard (1990) estimated 0.18 metres of surface peat was eroded from a palsa surface in Quebec in two successive winters. Wind erosion also occurs on mountain summits where heavily abraded and eroded remnants of blanket peat can be found. Luoto and Seppälä (2000) have noted these features on the fell summits in Finnish Lapland. Similar summit type erosion is also observed in the UK (Mackay and Tallis 1996). However, perhaps the most dramatic evidence for wind abrasion of soil is from Iceland, where wind erosion has stripped the landscape of thousands of hectares of surface soil (including peat) since settlement 1,130 years ago (Arnalds 2000; Figure 6.2a). The geomorphic processes that have triggered catastrophic erosion and produced the characteristic Icelandic 'rofabards' or erosional escarpments are very similar to those acting on bare peat producing very similar landforms (Figure 6.2b and 2c).

Wind erosion processes combine to produce a distinctive set of geomorphic landforms in peat (Figure 6.3). These range in scale from summit deflation flats and stream-lined hagg fields (Figures 6.3a and 6.3b), through mesoscale forms such as micro terraces and rofabards (Figure 6.3c and 6.3e), to microforms consisting of peat ripples (Figure 6.3d) and crust forms (Figure 6.3f). Through studies like those described in this chapter we are starting to understand how these features are formed and the significance of wind in their origin.

6.3   Mechanisms and Processes of Wind Erosion

Wind erosion of upland peat is dominated by direct shear from wind or by the impact of wind-driven rain (Figure 6.4). Understanding of the local air flow at the peat surface can be used to calculate the friction velocity which is often used as a predictor of sediment flux by wind (Campbell et al. 2002; Warburton 2003). This is calculated from the wind velocity profile using a semi-logarithmic equation:

$$\frac{u_z}{u_*} = \frac{1}{\kappa \ln(z/z_0)}$$ (eq. 6.1)

(a)

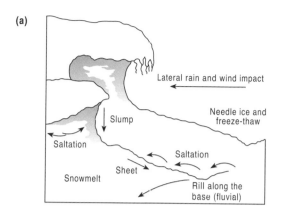

Lateral rain and wind impact

Needle ice and
freeze-thaw

Slump

Saltation

Saltation

Sheet

Snowmelt

Rill along the
base (fluvial)

(b)

(c)

Figure 6.2 Rofabard erosion features showing (a) schematic diagram of main processes, (b) example of classic rofabard from central Iceland and (c) peat rofabard from the North Pennines, UK

Figure 6.3   The range of peatland wind eroded landforms. (a) Deflation flat. (b) Wind-hagg fields. (c) Micro terraces. (d) Wind ripples. (e) Peat rofabard. (f) Deflation crusts. These examples show the range of scales involved and the combination of wet and dry wind-driven processes responsible for formation

where u_z = wind velocity at height z (m), u_* = friction velocity $(\mathrm{m\,s^{-1}})$, κ = von Karman's constant, z = height (m), z_0 = roughness length (m).

The assumption of a semi-logarithmic profile is complicated in the field by turbulence around topographic elements and, in such cases, a semi-logarithmic wind profile is unlikely to hold (Livingstone and Warren 1996). Given these assumptions it is therefore not surprising that

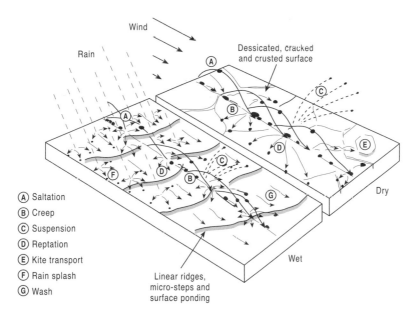

Wind

Rain

Dessicated, cracked
and crusted surface

(A) Saltation
(B) Creep
(C) Suspension
(D) Reptation
(E) Kite transport
(F) Rain splash
(G) Wash

Dry

Wet

Linear ridges,
micro-steps and
surface ponding

Figure 6.4   Schematic diagram showing differences in aeolian transport processes in dry and wet conditions. In dry conditions aeolian transport of dry peat particles and surface crust fragments dominates. In the wet and windy conditions aeolian transport still occurs but rain splash and wash are also important

Warburton (2003) found no consistent relationship between friction velocity and measured peat sediment flux.

In peatland environments it is important to distinguish between aeolian processes operating in dry conditions and those occurring during wet periods. Figure 6.4 shows the main modes of particle transport under dry surface conditions when the peat surface is often desiccated, crusted and cracked, and wet conditions where the peat is saturated and there is locally ponded water. In the dry, 'normal' aeolian processes dominate particle transport. Saltation (Bagnold 1941; Nickling 1988), suspension and creep (Wilson and Cooke 1980) actively occur and collisions of saltating particles with the surface dislodge particles causing either creep or reptation (a low-hopping motion). Occasionally sections of the dry surface peat crust may be entrained and lifted briefly on the wind in a process termed kite transport (Figure 6.4). Much of the eroded material is trapped locally in vegetation but some will occasionally form small bedforms including ripples. Finer material and dust are transported off-site and can enter higher level circulation. Under wet conditions the aeolian processes of saltation, suspension and creep still operate but these are augmented by rainsplash and surface wash. Rainsplash is particularly effective in detaching low-density peat particles which are readily transported by both the wind and surface wash (Figure 6.4). The combination of these processes results in a mass flux of eroded material in the direction of the prevailing wind resulting in microrelief features such as linear ridges and steps (Figure 6.4).

In upland Britain erosion by wind-driven rain is relatively common because of the propensity for high winds accompanying heavy rainfall. Under wet and windy conditions sediment entrainment is initiated by oblique rainfall impacts and sediment transport proceeds in a downwind direction under the influence of wind-driven rain (Moeyersons 1983; de Lima 1989; de Lima et al. 1992). In this situation, windward exposures experience more intense ballistic impacts and lift resulting in increased erosivity (de Lima 1992), and will also be subject to greater raindrop impact pressures from increased raindrop velocity and drop size (Erpul et al. 1998, 2004). Wind-splash is therefore the process in which wind and rain combine to cause erosion (Erpul et al. 2002b). There is increasing recognition that wind-splash processes are geomorphologically important in conditioning soil erosion (Vieira et al. 2004), and laboratory investigations have demonstrated that rain accelerated by wind increases erosivity on windward slopes (Erpul et al. 2002a and b, 2004; Pedersen and Hasholt 1995).

Invoking a wind-driven rain hypothesis explains neatly why the windward to leeward flux-density ratios reported by Warburton (2003) were 2–12 times those of leeward traps, which is remarkably consistent with the range of values reported by Foulds and Warburton (2007b) who measured ratios of 2–13. Other studies have also noticed a relationship between aspect and sediment yield. Francis (1990) reported significant differences in flux between N and SW orientations and E and S orientations in Mid-Wales. However, results were collected during drought and cannot be directly compared to splash-dominated processes. Birnie (1993) reported maximum erosion on northerly aspects in Shetland but related this to greater frost frequency rather than wind-driven processes. An excellent example of the direct link between aeolain process and form is Figure 6.5 which shows the relationship between the dominant wind pattern and orientation of streamlined landforms (haggs and mounds) at Moss Flats in the North Pennines. There is a strong association between the prevailing wind direction (mean direction 240°: vector strength 0.356) and the dominant orientation of streamlined erosion forms (mean direction 240°: vector strength 0.963) (Figure 6.5). Partly vegetated, streamlined hummocks and mounds are preferentially oriented towards the prevailing wind. Greater erosivity on windward slopes explains how these landforms come to be orientated to face the direction of prevailing weather

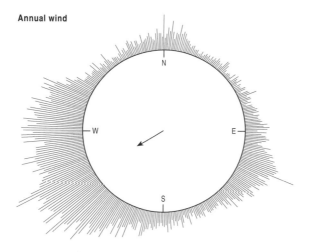

Annual wind

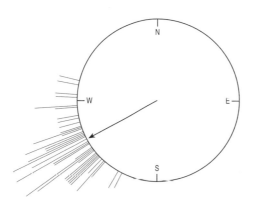

Orientation of streamlined erosion forms

Figure 6.5   The relationship between the prevailing wind direction and the dominant orientation of streamlined erosion forms, Moss Flats, Northern Pennines. Partly vegetated, streamlined hummocks and mounds are preferentially oriented towards the prevailing wind

systems. This is at odds with Tufnell (1969), who commented that wind erosion produces no lasting geomorphological imprint on the landscape. These conclusions apply to bare areas of relatively flat peat. On sloping ground it is expected that fluvial process and surface water erosion will dominate.

Under dry conditions fine dust particles are transported in suspension at much greater heights and over far greater distances than under wind-driven rain. Furthermore, under prolonged periods of dry weather, peat flats behave similarly to lowland soils and may develop intensive 'blows' (Pollard and Miller 1968). Wind erosion can occur during the summer when 'peat dust' is blown from the soil surface, particularly from areas of disturbance such as intensively grazed tracts. The quantitative import-ance of dust transport under dry conditions is uncertain due to the fact that very fine particles are transported at heights above the surface at a considerably greater height range than that sampled by the mass flux col-lectors. Locally, the soil crust may be dried to such an extent that once detached large peds, several centimetres in dimension, can be moved short distances by wind. The detachment of the surface crust occurs when the upturned edges of a cracked and desiccated peat surface protrudes above the general level of the ground surface (Figure 6.3f). These edges act as foci for local wind shear (Selkirk and Saffigna 1999), which are caught by the wind and entrained. This creates complex patterns of local surface roughness and which in turn affect surface stability, as noted by Campbell et al. (2002). Small sections of crust of light mass were seen to roll across the bare surface, sometimes moving by saltation, and were deposited close to the bare peat edge/vegetation interface. Elsewhere, large-scale crust stripping has been observed where a linear edge is exposed to the dominant wind direction and sections of crust are system-atically stripped from the receding margin as the crust edge progressively retreats (Tallis 1964a).

Finally it should also be noted that aeolian transport also occurs in winter when the surface crust is desiccated by ground freezing and can become detached. Often after particularly severe winters peat can be seen deposited on snow banks.

6.4   Direct Measurements of Wind Erosion of Peat

Identifying the dominant wind erosion processes in the field is compli-cated because modes of transport continually change (Wilson and Cooke 1980) and without direct intensive monitoring relative process rates cannot be determined.

Warburton (2003) reported the first direct measurements of the sig-nificance of wind action in the erosion of upland peat. Wind erosion monitoring was undertaken at Moor House in the North Pennines on a 3-hectare area of relatively flat, sparsely vegetated peat. Measurements using arrays of passive horizontal mass flux gauges (fixed orientation vertical slot gauges) together with a vertical array of mass flux samplers (directional) provided estimates of sediment flux (Figure 6.6a and 6.6b). A micrometeorological station recorded local wind-speed (four heights),

Figure 6.6 Examples of (a) the passive horizontal flux gauge, (b) Big Spring Number Eight (BSNE) horizontal flux sampler and (c) ring configuration of passive horizontal flux gauges

(a)

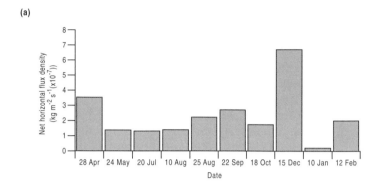

(b)

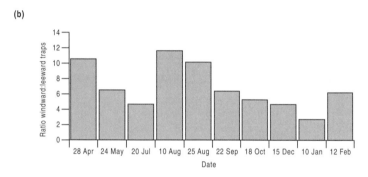

Figure 6.7   Warburton (2003) wind flux diagram of windward and leeward fluxes. (a) Mean net horizontal flux of eroded peat measured using the passive horizontal flux gauges. The period of observation extends from April 2000 to February 2001. (b) Comparison of windward and leeward mass sediment flux sampled using the slot traps. Windward sediment fluxes are approximately three to two times the leeward flux

wind direction, rainfall, and soil moisture and temperature conditions. These results are significant because they quantify for the first time the significance of wind action in the erosion of peat in a UK upland environment.

Using the flux samplers it has been possible to demonstrate that the horizontal mass flux of eroded peat in the direction of the prevailing wind can be up to 12 times greater than in the opposing direction (Figure 6.7). Analysis of the available wind velocity series has shown significant periods of windiness corresponding to individual windstorms whose severity and frequency increases in the winter months. Local micrometeorological data were used to try to predict sediment yields. However, it has not been possible to isolate the exact hydrometeorological controls which govern

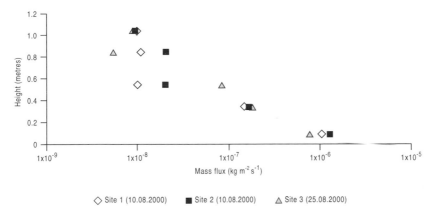

Figure 6.8   Vertical wind velocity profiles and flux rates (Warburton 2003). Examples of typical vertical mass flux density profiles. The three profiles are all measured at Moss Flats: Site 1, 10 August 2000; Site 2, 10 August and Site 3, 25 August 2000

wind erosion transport and the threshold of entrainment of particles from peat surfaces. Correlations between time-averaged friction velocity measurements, surface conditions and sediment flux did not reveal any consistent patterns in transport.

Warburton (2003) used BSNE samplers (Figure 6.6b) to measure the time-averaged horizontal sediment flux at six heights above the ground surface. Results (Figure 6.8) show that the peat mass flux with height declines dramatically with the majority of peat being transported within 0.3 metres of the ground surface. Significant horizontal fluxes of peat occur in both wet and dry periods. Under dry conditions the distribution will probably be different, reflecting greater mixing of lower density dry peat higher in the velocity profile. Significant horizontal fluxes of peat occur in both wet and dry periods.

Warburton (2003) has shown the dominance of windward peat fluxes over leeward fluxes (Figure 6.7); however the methodology used by Warburton (2003) meant that peat fluxes were only recorded at 240° (direction of the prevailing wind) and at 80° directly leeward of this. Wind direction can, however, be highly variable and strongly modified by local topography in upland environments (Rudberg 1969). As an alternative Foulds and Warburton (2007a and b) used an array of 16 mass flux samplers (Hall et al. 1994), arranged in a 3 metre-radius circle, to collect peat eroded by wind-assisted splash processes at intervals of 22.5° around a 360° circle. This allowed a full appreciation of the spatial nature of wind erosion dynamics from 16 cardinal compass directions (Figure 6.6c).

Foulds and Warburton (2007a) monitored wind-splash processes over a three-month period at Moss Flats in the North Pennines, northern England. Results from a circular arrangement of mass flux samplers linked to meteorological data analysis were used to assess spatial patterns of wind-splash erosion (Figure 6.9). Maximum peat flux was recorded between south–southwest and west–northwest orientations in association with low intensity frontal rainfall (typically 2–6 mm hr^{-1}). Although wind-splash processes were found to operate in any direction due to changeable synoptic weather patterns, windward fluxes from the direction of the prevailing wind (averaged over two- to three-week measurement periods) were generally 2–13 times the leeward (Figure 6.9).

The wind tunnel experiments of Campbell et al. (2001) comparing the erodibility of loose peat and crusted peat clearly demonstrate the importance of crust in controlling erosion (Figure 6.10). In these experiments the eroded volume of peat increased with increasing wind speed but the degree of humification of the peat (von Post values 3 to 6) did not appear to affect threshold velocities of entrainment. However, comparing the crusted peat with the loose peat it was shown that eroded peat volumes were several orders of magnitude less with the crusted peat (Figure 6.10). However these results represent only the initial stage of crusting, prior to desiccation. With increased desiccation cracks will form resulting in surface disturbance and increases in the local aerodynamic roughness.

6.5   Significance of Dry Conditions and Drought for Wind Erosion

Foulds and Warburton (2007a) found that peak flux during dry conditions was lower than under rainfall. This clearly demonstrates that rain enhances the effectiveness of wind in transferring sediment near to the peat surface (Erpul et al. 2002a and b, 2004). Low intensity wind-driven rain fluxes were an order of magnitude greater than typical dust flux rates recorded during a dry period and two orders of magnitude greater (up to 4.2×10^{-7} kg m^{-2} s^{-1}) during periods of sustained wet weather. Low intensity rainfall during periods of surface desiccation also produces significant amounts of wind-splash erosion. Terry and Shakesby (1993) reported that under simulated drought conditions in the laboratory, soil particles became hydrophobic, producing large lightweight sediment particles which were readily transported by rainsplash. Similar processes may well operate with peat soils because the hydrophobic nature of organic peat following drought has been widely reported (Holden and Burt 2002b).

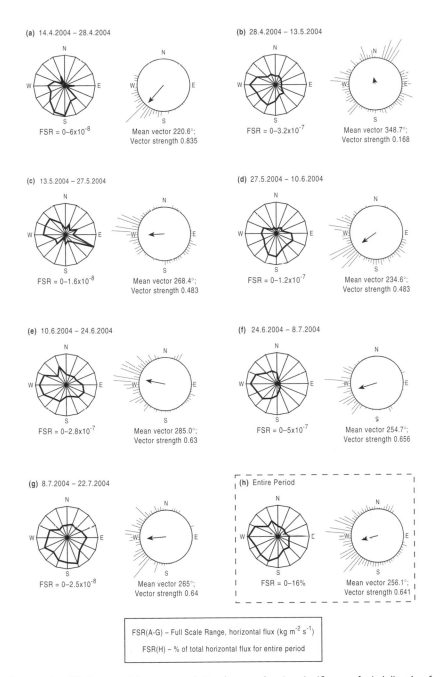

(a) 14.4.2004 – 28.4.2004

N
W — E
S

FSR = 0–6x10⁻⁸

N
W — E
S

Mean vector 220.6°;
Vector strength 0.835

(b) 28.4.2004 – 13.5.2004

N
W — E
S

FSR = 0–3.2x10⁻⁷

N
W — E
S

Mean vector 348.7°;
Vector strength 0.168

(c) 13.5.2004 – 27.5.2004

N
W — E
S

FSR = 0–1.6x10⁻⁸

N
W — E
S

Mean vector 268.4°;
Vector strength 0.483

(d) 27.5.2004 – 10.6.2004

N
W — E
S

FSR = 0–1.2x10⁻⁷

N
W — E
S

Mean vector 234.6°;
Vector strength 0.483

(e) 10.6.2004 – 24.6.2004

N
W — E
S

FSR = 0–2.8x10⁻⁷

N
W — E
S

Mean vector 285.0°;
Vector strength 0.63

(f) 24.6.2004 – 8.7.2004

N
W — E
S

FSR = 0–5x10⁻⁷

N
W — E
S

Mean vector 254.7°;
Vector strength 0.656

(g) 8.7.2004 – 22.7.2004

N
W — E
S

FSR = 0–2.5x10⁻⁸

N
W — E
S

Mean vector 265°;
Vector strength 0.64

(h) Entire Period

N
W — E
S

FSR = 0–16%

N
W — E
S

Mean vector 256.1°;
Vector strength 0.641

FSR(A-G) – Full Scale Range, horizontal flux (kg m⁻² s⁻¹)

FSR(H) – % of total horizontal flux for entire period

Figure 6.9 Wind rose and flux rose correlation diagram showing significance of wind direction for sediment transport (Moss Flats, North Pennines, UK) (after Foulds and Warburton 2007a)

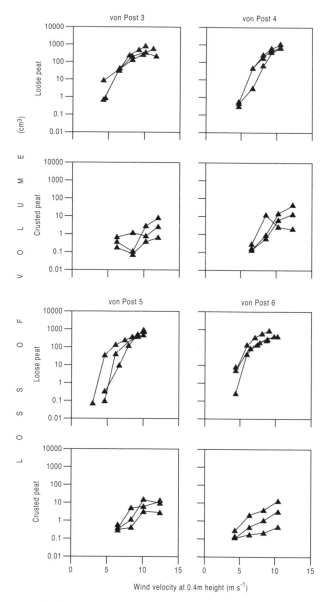

Figure 6.10 Loss of volume changes as a function of wind tunnel velocity for loose peats (top row) and crusted peats (bottom row), plotted against increasing degree of decomposition assessed using the von Post scale (after Campbell et al. 2002)

Erosion of blanket peat during drought was also studied by Francis (1990) in Mid-Wales. Francis reported that surface recession of bare peat faces was highly variable related to aspect, being greatest on southwest faces. It was concluded that peat shrinkage was responsible for maximum surface lowering and that wind erosion had minimal impact. However, it should be noted that Francis used erosion pins to monitor soil loss and whilst this method provides relatively accurate measurements of surface lowering, little insight into the processes is offered. Selkirk and Saffigna (1999) arrived at a similar conclusion studying wind and water erosion in a sub-Antarctic environment.

Hulme and Blyth (1985) observed that during a period of drought on Yell, Shetland, peat crusts could be observed blowing in the wind, but the amount of peat moved in this way is probably considerably less than that transported by water. However these observations relate to erosion in a gully system and this will differ according to the processes operating on a peat flat (cf. Bower 1960). Rates of wind erosion measured by Evans and Warburton (2005) suggest wind erosion in gullies is more important than on bare peat flats (Chapter 4). It should be noted that this was a wet period with no significant surface desiccation.

Foulds and Warburton (2007b) describe direct measurements and observations of blanket peat erosion by wind action during a short period of drought (surface desiccation) at Moss Flats in the North Pennines. Their measurements define the short timescale over which surface drying, desiccation and intensive erosion can develop in blanket peat environments. During a two-week period of sustained dry weather in May 2004, peat was rapidly eroded through a combination of crust detachment, dust deflation and wind-splash.

6.6   Conclusions

The main aims of this chapter were to review past research on wind erosion dynamics in upland environments and quantify wind erosion and aeolian processes. It can be concluded that wind erosion on bare peat flats and summit areas is an important dynamic process which produces a distinct suite of small-scale landforms. Maximum erosion occurs on SSW to WNW aspects and this is related to the mechanics of erosion by wind-driven rain. Windward peat fluxes are 2 to 13 times those in the leeward. Over time, peat hags come to be orientated to the direction of prevailing weather.

It is clear that peat erosion by wind is an important process in upland peatland landscapes. However its impacts will only be noticed on flatter ground where fluvial processes do not dominate. Evans and Warburton

(2005) clearly demonstrate this in their peatland sediment budget; fluvial sediment fluxes are one to two orders of magnitude greater than aeolian fluxes. It is important to attach a caveat to this statement because the conditions for accelerated wind erosion on relatively flat expanses of bare peat, exposed to the combined elements of wind and rain, are widespread in some upland peat landscape where summits vegetation has been degraded through overgrazing (Mackay and Tallis 1996), or moorland burning has stripped the surface vegetation (Radley 1965; Maltby et al. 1990). The evidence base for assessing the importance of wind erosion in peatland soils is very slight.

Detailed event-based observations of erosion episodes are required to better define mechanisms and thresholds of entrainment for different types of peat under varying hydrometeorological conditions. It is clear from the work of Foulds and Warburton (2007a and b) that earlier work by Warburton (2003) was carried out during a period dominated by wind-driven rain. This is important because these data were used in the construction of the sediment budget of Evans and Warburton (2005; Chapter 4). This emphasizes the need for longer term studies of upland sediment dynamics so that the full range of potential erosion processes can be properly compared.

Chapter Seven

Peat Erosion Forms – From Landscape to Micro-Relief

7.1   Rationale and Introduction

Upland peatlands have developed a characteristic range of geomorphic features by virtue of their steep relief, harsh climate and the dominance of peat as the surficial soil type (Radforth 1962). As outlined in Chapter 1 this has led to a distinctive suite of landforms which collectively define the upland peat landsystem (Figure 7.1). A landsystem can be defined as a set of genetically related landforms and surficial deposits, formed in a particular environmental setting. Therefore, in an upland environment key factors such as slope and topography, climate and drainage and vegetation will determine rates of geomorphic processes, which in turn will give rise to a suite of landforms and deposits characteristic of that environment.

At the smaller scale the special properties of peat are very important in governing the types and rates of geomorphic processes that operate. For example, the low density of both wet and dry peat is important in determining transfer rates by both fluvial and aeolian processes (Chapters 4 and 6). The unique structural and geotechnical properties of peat control its hydraulic behaviour and the stability of the peat mass on a hillslope (Chapters 2 and 5). The special water relations found in peat also determine the response of upland environments to hydrological extremes, whether they be floods or drought.

The peat landsystem diagram outlined in Chapter 1 summarizes the main geomorphic features present in an upland temperate peat landscape (Figure 1.8). It is clear from this simple diagram that the main geomorphic forms can be grouped in accordance with the main forces involved in their formation and the local slope and geomorphic setting. For example, linear gullies are found on steep slopes, whereas pool complexes

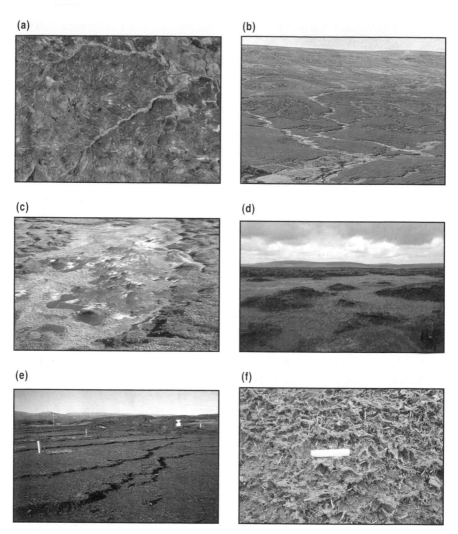

Figure 7.1 A nested hierarchy of peatland geomorphic forms. The example shown is from Moor House in the North Pennines and attention is focussed on a small peat flat (Moss Flats). Six representative scales are illustrated with the approximate linear dimensions of each (m). (a) The peatland drainage basin – 10^3. (b) The blanket bog slope catena incorporating dissection and bare ground erosion – 10^2. (c) A local slope element (peat flat) – 10^1. (d) Peat haggs, pools and mounds – 10^0. (e) Small-scale erosion and depositional features – 10^{-1}. (f) surface microtopography – 10^{-2}

occur where the drainage is impeded on flats or in topographic hollows (Figure 7.1). This type of organization in the landscape arises due to interaction of processes over a range of scales. The classification of peat-land geomorphic forms can be characterized at three basic scales:

- *Macroscale* (region/catchment scale) – regional topography exerts a strong influence on blanket peat thickness and drainage patterns. The range of slope angles in a particular area to some extent controls the accumulation depths of peat and this is closely related to the ecohydrol-ogy of the peatland which cannot be separated from the prevailing regional climate (water balance).
- *Mesoscale* (slope-channel scale) – at a smaller scale the upland peat landscape can be divided into slope elements which have a character-istic arrangement of landforms. This often involves a peat-slope catena consisting of summit erosion complexes, mid-slope gully erosion and mass movements, and lower slope peat deposition and/or coupling with the mainstem channel network. This simple slope series contains the key elements that define the upland sediment cascade and as such can be viewed as the operational scale for sediment budget studies.
- *Microscale* (material structure scale) – at a smaller scale peatland sur-faces have a wide variety of smaller scale landforms which contribute to 'form' roughness (small landforms on the scale of m and dm) and 'surface' roughness (cm or mm scale). In this context the form rough-ness can be thought of as the haggs, mounds, pools and ridges that make up the peat landscape, whilst the surface roughness is really composed of vegetation or the microtopographic forms which develop on bare peat surfaces. At this scale the structural properties of the peat are important in governing the stability and morphology of surface forms. Surface roughness is the preferred term here because peat sur-faces do not have well-developed grain structures.

These scales are illustrated in Figure 7.1, which shows a nested hierarchy of peatland geomorphic forms. The example is from Moor House in the North Pennines with a specific target landform assemblage of a small peat flat (Moss Flats) (Warburton 2003). The diagram shows the peatland geomorphology representative at six different scales characterized by dif-ferent linear dimensions (m) over a range of six orders of magnitude. At the largest scale the peatland drainage basin (Figure 7.1a) is shown, within which are various slope systems each generally characterized by the blanket bog slope catena incorporating dissection and bare ground erosion on a small peat flat (Figures 7.1b and 7.1c). At progressively smaller scales the detail of the peatland surface is dominated by peat haggs, pools and mounds (Figure 7.1d) which add to the general surface roughness. At the smallest scale, local erosion and depositional features dominate (Figure 7.1e), resulting in a wide range of surface microtopography (Figure 7.1f). This simple example shows the nesting of the three main scales: macro (Figure 7.1a); meso (Figures 7.1b and 7.1c); and micro (Figures 7.1d, 7.1e and 7.1f). The aim of this chapter is to describe examples of the characteristic landforms that make up these broad-scale divisions and illustrate how these scales combine in a simple landsystem model to produce different peatland environments.

The structure of this chapter follows the hierarchy of scales identified above, beginning with the macroscale and working progressively to the microscale. This is followed by a brief consideration of how this proposed classification relates to the well established ecohydrological definition of ombrotrophic mires described by Lindsay (1995). This comparison is important because bridging the gap between geomorphological and ecological controls is an important goal. Given the preliminary nature of the proposed framework the chapter concludes with some general observations.

7.2   *Macroscale* – Region/Catchment Scale

The general peatland landsystem defined in Chapter 1 (Figure 1.8) provides a useful conceptual model of landform linkages in peatland environments. However such general models need refinement at the regional scale.

Figure 7.2 shows contrasting views of three UK upland peatlands in Northumbria, the North Pennines and South Pennines. Accompanying the photographs is a sketch diagram showing the altitude, general topography of the three areas and the approximate peat cover distribution. The most striking thing about this diagram is that the three areas have very distinctive topographic settings, being essentially convex, concave and linear in overall form. To some extent topography alone can explain the peat distribution but it is clearly evident the three areas have experienced very different peat-forming conditions both in response to environmental factors (e.g. rainfall and altitude) but also anthropogenic factors such as pollution histories and land-use (Clement 2005). Figure 7.3 is a summary of the distribution of peat erosion type with altitude for the three contrasting peatlands (cf. Tallis 1985b). Four erosional classes are distinguished: uneroded, linear, dendritic or anastomosing dissection. These are expressed as a percentage of the area of peat in a particular altitudinal band. A value of 10-metre altitude was used to define these areas. From this figure it is evident that Moor House in the North Pennines has the greatest proportion of uneroded peat across the greatest altitudinal range. Cheviot contains the most widespread evidence of dissection (Figure

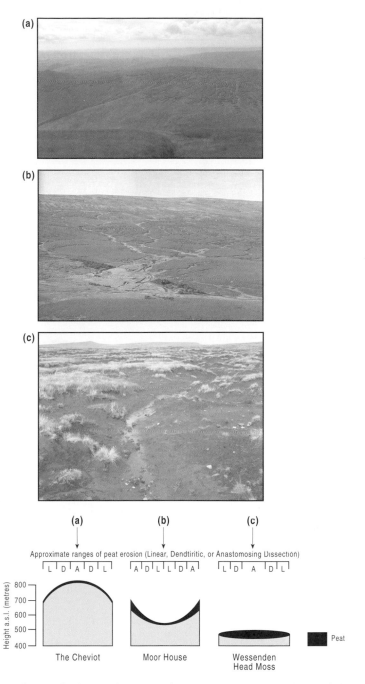

Figure 7.2 Photographs showing three views of contrasting UK upland peatlands: (a) Cheviot, Northumbria; (b) Moor House, North Pennines; (c) Wessenden Head Moss, South Pennines. Accompanying the photographs is a sketch diagram showing the altitude, general topography of the area and peat cover distribution. Approximate ranges of peat dissection are shown (based on Clement 2005)

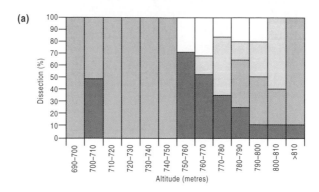

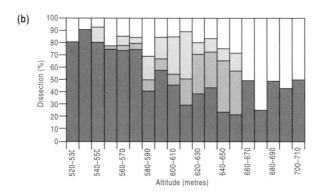

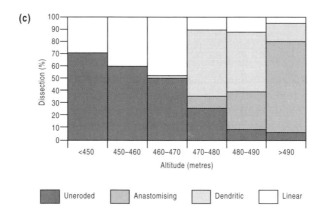

Figure 7.3 Diagram showing the distribution of erosion type with altitude. Four erosional classes are distinguished (uneroded, linear, dendritic or anastomosing dissection), expressed as a percentage of the area. (a) The Cheviot (based on Wishart and Warburton 2002). (b) Moor House (based on Clement 2005). (c) Wessenden Head Moss (based on Clement 2005)

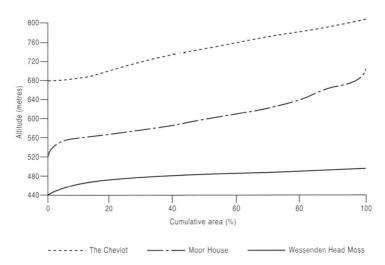

Figure 7.4   Hypsometric curves for (a) Cheviot, Northumbria; (b) Moor House, North Pennines; (c) Wessenden Head Moss, South Pennines (based on Clement 2005)

7.3a). In the South Pennines the area of uneroded peat decreases progressively with altitude. In general terms 52% of the peat blanket is still intact at Moor House, 36% in the South Pennines and only 20% in the Cheviot (Clement 2005). Patterns of dissection vary markedly between the three areas. In the North Pennines dissection is strongly linear whilst on the Cheviot anastomosing systems dominate. The South Pennine site shows an approximate equal distribution of all types of dissection (Figure 7.3). The explanation for these differences can to some extent be found by examining the hypsometric curves for the three study areas (Figure 7.4). The flat convex curve characterizing Wessenden Head Moss clearly shows a relatively flat landscape with steepening slopes with distance downslope – ideal conditions for a smooth transition between dissection types. At Moor House (Figure 7.4) mid- to upper slopes are much steeper and in these areas linear dissection is well developed; areas of flatter ground are not particularly common. On the Cheviot the plot is fairly linear and has some similarities with Moor House, except that the overall profile is more dome-shaped hence linear dissection is less likely.

Another factor illustrated in Figure 7.4 is the altitudinal differentiation between the three study regions; this will undoubtedly have a bearing on the type of geomorphic processes and cover types dominating at the various altitudes and will condition the form and nature of erosion. Care needs to be exercised when interpreting these results because the proportions of erosion found in a particular altitudinal band relates to the absolute amount of peat within that zone.

7.3   *Mesoscale* – Slope Catena Scale

The idea of a systematic arrangement of geomorphic processes and forms along a slope catena is implicit in many of the early models of blanket peat geomorphic systems (e.g. Bower 1960; Radley 1962; Barnes 1963; Chapter 4, Figure 4.4). However the significance of such models was neither fully defined nor reconciled amongst early workers because of two main problems. Firstly, peatland erosion processes were not adequately quantified; hence arguments about the relative importance of, for example, wind erosion over fluvial processes could not be fully resolved. Secondly, regional differences in upland peat geomorphic systems were not fully recognized. Based on the evidence presented here and in other parts of this book we can generalize about the operation and dominance of these processes and therefore propose a simple model.

The blanket peat landscape can be conceptualized in terms of a simple convex-linear-concave slope series (Figure 7.5) (McGreal and Larmour 1979). On exposed summits and interfluves peat is open to weathering and erosion processes and commonly consists of degraded hagg fields with runoff being fed to lower parts of the landscape (as indicated in the meso-tope model of Lindsay [1995]). At the foot of the slope, water accumulates and results in areas of deeper peat with associated pond complexes. The

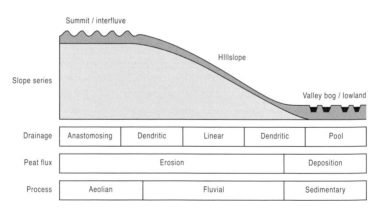

Figure 7.5   The upland blanket bog geomorphic slope catena showing the approximate limits of surface drainage characteristics, erosional status and process zones. Intensity of process will vary between peatlands

intervening slopes are commonly linear and will be dissected by gullies depending on the local runoff conditions and the integrity of the hillslope vegetation. Drainage is an important component of the hillslope. In flat summit areas ephemeral anastomosing drainage patterns may develop connecting local areas of ponding water. These translate into more distinct dendritic patterns as slope angle increases. Further steepening of the slope results in linear gullies which have the potential to deeply incise the peat blanket. Towards the foot of the slope dendritic or divided drainage may also occur before the surficial drainage is subsumed with the deeper peat and pool complexes at the slope foot.

Differences in slope process regimes lead to characteristic patterns of erosion and deposition with erosion being much more important on the exposed hillcrests and active hillslopes (Mackay and Tallis 1996). Low-lying areas tend to accumulate eroded material if the hillslope drainage does not drain directly into a mainstem river. The key geomorphic processes operating within these landscape elements are spatially differentiated. On the flat exposed hill summits aeolian processes, particularly wind-driven rain, are of greater significance than fluvial processes. However, as slope angle increases fluvial processes rapidly dominate the nature of sediment transfer. Sedimentation of eroded hillslope peat in the valley bog depends on the complex interactions between flow and vegetation. Vegetated pool complexes are very efficient sediment traps even for low-density eroded peat.

When considering the geomorphology of peat landscapes the significance of pools and pool complexes is often overlooked because the focus tends to be on erosional forms. Although pools are sites of low erosional activity they are extremely important as depositional sites. Erosion can quickly destabilize pool complexes leading to rapid drainage and loss of peat. Following disturbance, water-tables fall and the upper peat can become desiccated and oxidized, and plant complexes wither leaving the bog surface prone to further erosional forces.

The mesoscale is often seen as the 'relevant' scale for assessing the significance of erosional fluxes in sediment budgets because it is the scale most appropriate for management. Sediment budgets also provide a framework for assessing the geomorphic status of a particular peatland system. By assessing the rates at which processes are operating and the linkages between specific transfers and storage elements the general 'connectivity' of the sediment cascade can be established and rates of landscape change inferred (cf. Chapter 4). The significance of peat erosion can be easily contextualized at the small catchment or hillslope scale and these changes in the upland system result in important consequences lower down the catchment. However, sediment budgets alone do not inform us of the cause of erosion or the processes that control rates of

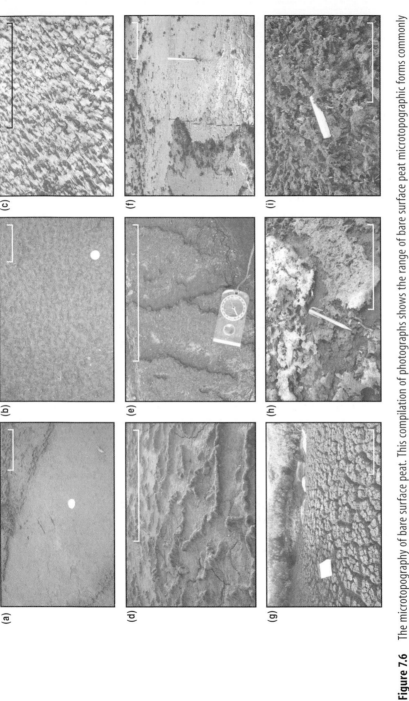

Figure 7.6 The microtopography of bare surface peat. This compilation of photographs shows the range of bare surface peat microtopographic forms commonly found on upland blanket bog in the UK. The scale bar shown in each picture is 0.3 metres. (a) Smooth surface of redeposited peat. (b) Pitted surface typical of vertical rainfall. (c) Toothed peat usually associated with wind-driven rain. (d) Linear ridges and grooves formed under wind-driven rain and surface wash. (e) Ripple-steps (wave-forms) usually developed due to a strong lateral sediment flux. (f) Major step or wash-front advancing from left to right and often associated with smaller features (d and e). (g) Surface desiccation cracking (may form both in summer and winter). (h) Crust formation sometimes associated with surface algae. (i) Nubbins or frost-fluff caused by surface frost heave of the bare peat surface. The form of these features will vary dependent on the nature of the peat material and the magnitude and frequency of the dominant geomorphic processes sculpting the surface

erosion, they only quantify the main fluxes. Smaller scale understanding (e.g. the hillslope scale) of key hydrological and geomorphic processes is still required, for example gully erosion (Chapter 3 and 4), peat mass movements (Chapter 5) and wind erosion (Chapter 6).

7.4   *Microscale* – Material Structure Scale

A key material characteristic of peat which has been emphasized in this book is its very low density as a geological material. This makes peat highly susceptible to the forces of erosion particularly by wind and rain. In addition a high water-content also imbues important properties in relation to wetting and drying cycles and frost heave.

In combination these forces produce a highly dynamic media which undergoes rapid deformation at the surface. The degree to which the peat will deform and the microtopography that results are not only conditioned by extrinsic factors (cf. Chapter 3), they are also strongly controlled by the intrinsic properties of the peat. This is significant because peat is a highly variable material in terms of its composition and this is of major importance in microtopography studies because the composition of peat macrofossils can form important structural elements in micro-landforms at the surface. Furthermore vegetation plays an extremely important role in providing structure to microscale landforms; many pool and hummock complexes are developed because of the small-scale ecology.

Bower (1959) was one of the first geomorphologists to recognize small-scale geomorphic features of eroding peat landscapes and classify them in accordance with the main governing processes. Table 7.1 describes the main agents of peat erosion and the characteristic surface forms that result. This systematic genetic classification of peatland microforms is useful inasmuch as it provides clues to the origin of certain microforms. Bower (1959) emphasized the importance of four main agents of erosion: running water, gravity, heavy rain/hail and freeze-thaw activity (Table 7.1), however, many of these processes are also strongly influenced by wind action. Given the equifinality in some of the small-scale landforms that result and the difficulties in assigning particular genetic origins this scheme has not really been adopted by other geomorphologists.

An alternative to this approach is the simple pictorial key of bare peat microforms presented in Figure 7.6. This compilation of photographs shows the range of bare-surface-peat microtopographic forms commonly found on upland blanket bog in the UK. The organization of the images is deliberate and serves as a useful preliminary classification scheme. Photographs (a) to (c) show increasing roughness of single-element micro-forms (pits and teeth). These are often associated with increasing energy

Table 7.1   The Bower (1959) approach to classification of peat geomorphic form — process relations. The table shows the basic forms and the dominant agents of erosion associated with them. Examples of many of these are shown in Figure 7.6

Agent of erosion	Process of erosion	Characteristic peat surface features
1 Running water	Sheet wash	Horizontally aligned wave-forms
		Smoothing of toothed surfaces
	Semi-channelled sheet wash	Diagonal and vertical wave forms
	Runnelling (Rilling)	Minor gullying of weathered peat
2 Gravity & heavy rain	Flow downslope of semi-liquid surface peat during and after heavy rain (micro-debris flows)	Terracettes in the weathered peat – ridging
	Gravity & rain & wind — Flow of semi-liquid peat on peat flats, direction of flow due to wind direction	Stepped surface on peat flats
	Gravity alone — Large scale slumping of loose, weathered peat on gully sides	Enlarged talus at the foot of peat faces
3 Heavy rain, hail (no wind)	Displacement of peat due to impact of large rain drops and hail falling on the surface	Minor pitting of the surface of weathered, primary peat and redistributed peat
	Heavy rain, hail, strongly driven wind — Rain and hail driven forcibly across the surface	Grooves
4 Freeze-thaw & gravity	Solifluction movement of weathered peat down gully sides	Enlarged talus

of wind-driven rain (Figure 7.6 a–c). The next set of three images show linear features whose roughness height generally increases from (d) to (f) in response to the magnitude of the lateral material flux (Figure 7.6 d–f). The final three pictures show three miscellaneous microforms associated with: surface desiccation cracking, which may form both in summer and winter months; crusting associated with surface algal growth, typical on slowly drying surfaces; and 'nubbins' or frost-fluff caused by surface frost heave of the bare peat surface. These features are significant for three principle reasons. Firstly, they increase surface roughness which has important feedback on near-surface sediment transfer mechanisms. Secondly, they indicate the relative magnitude and direction of sediment transfer. Thirdly, they provide a mechanism by which surface peat is detached from the main peat mass and promote local soil loss (Chapter 3). The form of these features will vary dependent on the nature of the peat material and the magnitude and frequency of the dominant geomorphic processes sculpting the surface. Microforms can show enormous variability over the scale of a few metres, for example the face and toe of a peat hagg or the opposing aspects of an incised gully. Such differences are controlled by differences in local microclimate which influence weathering and erosion processes; however, the exact nature of these controls has not been quantified.

7.5 Linking the Geomorphological and the Ecohydrological

The focus of this book is on upland peat erosion and as such the dominant mire type under consideration is the ombrotrophic mire whose wetland origin and nutrient status is derived from direct atmospheric precipitation alone (Lindsay 1995). Classic models of bog development do not always apply in areas of blanket bog where the upland topography exerts a strong influence on local peat-forming conditions. In this setting 'paludification' (peat formation under water-logged conditions directly above impermeable soil or rock) is a dominant mechanism until the peat is a sufficient depth that peat domes form. It has been proposed that raised bogs will grow until they reach a limiting profile whose form and maximum height can be predicted in accordance with the so-called groundwater mound theory. Such forms may be approximated in upland peatland landscapes but they are rarely fully attained due to the complexities of the local topographic environment (Lindsay 1995). Under such conditions bog complexes evolve in response to local hydrological gradients which may eventually coalesce into complex forms. It is important to also remember that such development is punctuated by periods of drought and episodes of erosion which severely disrupt these simple bog-development models. Erosion in particular may produce long-term changes in the peat mass due to intense dissection and lateral striping of peat. For example, where the peat blanket has been severely degraded mire complexes become fragmented and the hydrological functioning can be altered over periods of millennia.

As mentioned above, regional differences in bog type and surface expression arise due to local differences in both climate and topography

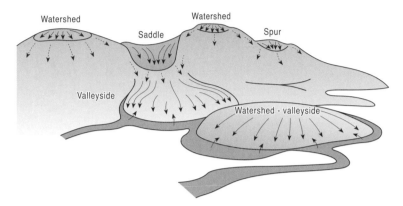

Figure 7.7   Hydromorphological and topographical characteristics of upland blanket bog (from Lindsay 1995). Major mesotopes are shown along with the dominant water flow lines

(Ivanov 1981; Lindsay et al. 1985; Figure 1.1). In upland peatlands complex arrangements of the basic mesotypes occur but they are linked into a continuous peatland system by the continuity of the peat blanket, for example watershed bog may grade laterally into a valleyside mire (Figure 7.7). At a smaller scale, bog surface microtope patterns vary according to the local hydrological conditions and dominant vegetation assemblages. Associated with this a number of structural mire microforms occur on ombrotrophic peat bog in upland Britain which Lindsay (1995) divides into two main classes; terrestrial and aquatic.

In the context of this chapter, and the book as a whole, the scheme can be applied to upland blanket bog (ombrotrophic bog types). This is summarized in Table 7.2. If we consider this classification in relationship to the preceding discussion and in particular the simple model defined in Figure 7.5 it is possible to usefully marry together the general ecological concepts summarized in hydromorphic mesotopes (Lindsay 1995; Figure 7.7) with the geomorphic processes operating in the landscape. This is illustrated in Table 7.2 which shows the cross correlation between the Lindsay (1995) scheme and that which has been developed here. Even a cursory comparison of Figure 1.1b and Figure 7.7 clearly illustrates the commonality of the two schemes. Whilst matches can be found in Table 7.2, the classification of erosional gullies is problematic because ecologically they are considered a microtope but geomorphologically a mesoscale feature. These differences are probably reconciled in the sense that a gully as a longitudinal feature is a mesoscale landscape element but if considered in cross-section it is a microtope element. In addition several features

Table 7.2   Hierarchical classification of upland blanket peat landforms (Lindsay 1995)

Ecohydrological classification (Lindsay 1995)		Geomorphological classification
Scale	*Type*	
Macrotope	Blanket bog	Macroscale – region/catchment
Mesotope	Watershed	
	Spur	
	Valley side	Mesoscale – slope catena
	Watershed-valley side	
	Saddle	
	Eccentric	
Microtope (terrestrial)	Low ridge	
	High ridge	
	Hummock	
	Peat hagg (erosional)	
	Peat mound	Microscale
Microtope (aquatic)	Sphagnum hollow	
	Mud-bottom hollow	
	Drought-sensitive pool	
	Permanent pool	
	Erosion gullies	Mesoscale

recognized as central to the ecohydrology scheme are not clearly recognized in the geomorphic classification, and vice versa. This too is easily explained as the ecohydrology approach places emphasis on peat accumulation and ecological stability whereas the geomorphic schemes emphasizes sediment transport and erosion.

7.6   Conclusions

This chapter has developed a framework for considering how the various geomorphic forms associated with erosion of blanket peat combine to produce characteristic peatland landscapes (Radforth 1962). The chapter has been written from a geomorphological perspective, therefore it has a clear bias towards erosional processes. In the next chapter we have identified the importance of re-vegetation and sedimentation as key processes encouraging deposition and stabilization of upland peat areas (Bragg and Tallis 2001). Such processes depend intimately on the structure of local vegetation and the interaction of these ecological communities with surficial geomorphic processes. These key interactions between sediment dynamics and vegetation are explored in Chapter 8 in the context of peatland restoration and are a key area for future research.

Furthermore it is very important we recognize the differences that exist between different upland peatland systems at the regional scale. It is not sufficient to simply measure or estimate the long-term rate over which a process has operated; we must place that process rate in the wider context of the sediment budget. Warburton et al. (2003) strongly criticized the recent national survey of upland erosion in England and Wales on this basis. The framework presented here, and the recognition of regional variation in erosion type associated with hypsometry, provides a potential approach to upscaling representative regional sediment budget assessments to provide more soundly-based national estimates.

Chapter Eight

Sediment Dynamics, Vegetation and Landscape Change

8.1 Introduction

Tallis (1995) has argued that management of blanket mires is dependent on an understanding of the long-term dynamics of the peatland system. In particular he cites correct identification of the causes of peat erosion as critical. If peat erosion represents a natural end point to peat accumulation, because of the perceived physical instability of large depths of saturated peat (Conway 1954; Pearsall 1956; Ivanov 1981; Tallis 1985a; Foster et al. 1988), then the dramatic landforms of eroded mires represent an important microtope within the range of mire forms worthy of preservation in their own right (Lindsay et al. 1988; Tallis 1995). The alternative view is that the widespread peat erosion seen in upland Britain today is a result of intense human pressure on the ecosystem over the past 500 years and that the erosion represents an unnatural alteration of the pristine system requiring intervention to manage the negative consequences of accelerated erosion. The detailed work on the causes of erosion by Tallis in the Southern Pennines (Tallis 1985a and b, 1987, 1994; Tallis and Livett 1994; Tallis 1997b and c; Tallis 1998) has demonstrated that the onset of erosion has many causes; some erosion is natural and some is triggered by both direct and indirect forms of human agency as discussed in Chapter 1. This appears to support the argument for intervention and management of accelerated erosion. There is a second important dichotomy in our understanding of the functioning of eroding peatlands. The erosion of upland peat may be regarded as catastrophic, the result of an irreversible crossing of a threshold which shifts the system from actively growing mire to a situation of rapid and progressive erosion. The alternative to this is represented by the conceptual model of erosion and re-vegetation of peatlands presented in Chapter 4. This hypothesizes that

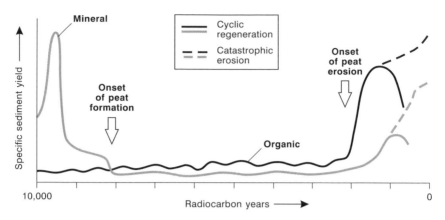

Figure 8.1   Conceptual model of Holocene variability in sediment yield from upland peatlands

erosion and regeneration of some sensitive upland mires is a natural process, so that the natural condition of mature blanket mire is to support a mosaic of intact, eroding, and re-vegetating peat. The aims of this chapter are to consider the implications of these alternative hypotheses for the nature of long-term sediment budgets in upland peatland catchments, to assess these hypotheses in the light of current evidence, and to consider the likely impacts of future climate change on long-term sediment dynamics.

8.2   The Effect of Peatland Dynamics on Long-Term Sediment Budgets

The development of extensive blanket peat cover across upland landscapes during the Holocene has had a major effect on the quantity and quality of sediment flux from upland systems. In wetter areas where peat soils are the natural edaphic climax and peat formation began in the early Holocene the landscape preserved below the peat is a deglacial and periglacial surface. In many other areas where peat formation was delayed until the mid-Holocene there are preserved human landscapes at the base of the peats suggestive of an anthropogenic influence on the initiation of peat formation (Caulfield 1983). Figure 8.1 considers the likely pattern of Holocene sediment yields from mineral and organic surfaces in relation to the development of blanket peat cover. The mineral sediment yield from this cool and unstable paraglacial (Church and Ryder 1972) environment would have been considerably above contemporary rates. At the

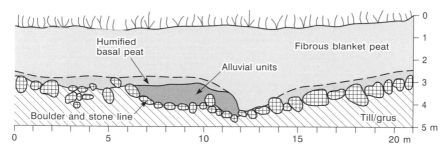

Figure 8.2   Alluvial valley sequence overgrown by blanket peat in Co. Wicklow, Ireland (after Thorp and Glanville 2003)

onset of peat formation paraglacial sediment yields are likely to have been significantly curtailed by the spread of a stable moorland vegetation cover across the upland landscape. At the same time the increase in vegetation cover would slightly increase organic yields. The second major perturbation in the Holocene sediment flux of these upland systems occurs with the onset of peat erosion. This produces significantly elevated levels of organic sediment yield but also in advanced stages re-exposes the buried periglacial landscape reactivating the paraglacial sediment store and increasing mineral sediment yields. Projecting current rates forward we can consider two possible scenarios (Figure 8.1). The first assumes catastrophic and irreversible peat erosion and shows a continued rise in both mineral and organic yields, whilst the second assumes cyclic regeneration of eroded surfaces such that the sediment yields decline.

Some interesting evidence for the effects of blanket peat spread on fluvial sediment systems comes from recent work by Thorp and Glanville (2003). They report the complete burial of a first-order stream valley system by blanket peat in Co. Wicklow, Ireland. The timing in this case study varies from the model presented above in that complete burial was not achieved until 2,200 radiocarbon years BP but significant deposition of alluvial material (dating to the mid-Holocene) suggests that prior to this the active fluvial channel was delivering large quantities of sediment downstream (Figure 8.2). Similar burial of pre-peat drainage lines has been reported from Axel Heiberg Island by Vardy et al. (2000) and from North Wales by Hall and Cratchley (2005). Mol (1997) suggests that significant peat development in the Niederlausitz during the early Weichselian was associated with reduced fluvial activity. At the landscape scale the only period with clearly documented evidence of widespread peat erosion is during the last millennium. However, locally reactivation of the mineral sediment store beneath the peat may have occurred periodically throughout the Holocene. For example, Ashmore et al. (2000) document significant mineral inputs to mire surfaces in the Hebrides caused by episodic mass movement of upslope peats between 3,000 and 1,750 BP. Therefore whilst Figure 8.1 might be a reasonable representation of patterns of Holocene sediment yield in the uplands generally, the specific local pattern may vary significantly.

8.3   Re-Vegetation of Eroding Peatlands

Central to the question as to the extent to which contemporary peat erosion represents catastrophic degradation of upland mires or a temporary period of instability is the ability of eroded mires to re-vegetate and reinitiate peat growth. There are two sources of evidence on the propensity of bare peat surfaces to re-vegetate; studies which have attempted to artificially re-vegetate bare peat, and work on the spontaneous re-vegetation of mined or eroded peat surfaces. These are considered separately below.

8.3.1   Artificial re-vegetation of bare peat surfaces

Research into the artificial re-vegetation of bare peat has addressed re-vegetation of sub-aerially eroded peatlands and also restoration of landscapes where peat mining is important, particularly in Scandinavia, Canada and Ireland (e.g. Cooper et al. 2001; Farrell and Doyle 2003; Tuittila et al. 2003; Campeau et al. 2004).

The most extensive work on eroded moorland has been undertaken in relation to planned restoration of the severely eroded landscapes of the Southern Pennines (Tallis and Yalden 1983; Anderson et al. 1997). This work focussed on attempts to re-establish cover of *Calluna vulgaris* on bare and eroded ground. Greatest success was achieved where some initial surface cover protected the heather seed, either through a mulch or a nurse crop of grasses. Winter seedling mortality associated with frost disturbance of bare surfaces was a major limitation on re-vegetation and the mulch or nurse crop plays a role in insulating/stabilizing the surface. Removal of grazing animals also enhances rates of re-vegetation. This is consistent with the findings of Gore and Godfrey (1981) who demonstrated that the presence of sheep limited colonization in reseeding and fertilizing trials. Fertilizer additions played a role, particularly in promoting rapid growth of the nurse crop in the early years of re-vegetation, establishing a deep root system to counter the effects of surface instability in the winter. Richards et al. (1995) describe a series of experiments assessing the use of *Eriophorum angustifolium* transplants to re-vegetate bare peat in hostile conditions on the Kinder Scout Plateau in the Southern Pennines. This is of particular interest since the *Eriophorum angustifolium* is a common pioneer species in areas of spontaneous re-vegetation. They conclude that the major limiting factor is the acidity of the surface peat, and in common with the work of Tallis and Yalden the best cover was established after application of lime and fertilizer. The general conclusion of the Peak District work was that re-vegetation of all surfaces was possible given suitable interventions (Tallis and Yalden 1983).

In North America the focus of peatland re-vegetation has been restoration of peat mining sites. These peatlands are more typical of raised bog globally because, unlike the Southern Pennines, they have *Sphagnum* as a major component of the vegetation. Consequently the focus of re-vegetation work has been the establishment of *Sphagnum* as a principal peat-forming species. *Sphagnum* requires a high water-table so much of the effort in peat mine restoration has been in raising water-tables and providing suitable pool environments for *Sphagnum* growth (Jauhiainen et al. 2002; Price et al. 2002; Campeau et al. 2004). Price and Whitehead (2001) have identified maintenance of high water-tables year-round as essential for the re-colonization of peatlands by *Sphagnum*. Brief periods of lower summer water-table depress *Sphagnum* regeneration and the most favoured sites are topographically low water-gathering sites. In a series of studies Rochefort and co-workers (e.g. Campeau and Rochefort 1996; Rochefort and Bastien 1998; Rochefort 2000; Rochefort et al. 2003) have established the range of conditions necessary for successful re-vegetation with *Sphagnum*. Three conditions are central (Figure 8.3): rewetting, provision of mulch, and spore availability. Comparison with the work in the UK is interesting since Tallis and Yalden (1983) similarly identify seed provision and some initial protection (mulch or nurse crop) as key to re-vegetating deep peat.

8.3.2   Natural re-vegetation of eroded landscapes

Forty-nine per cent of major erosion features surveyed in upland England and Wales by McHugh et al. (2002) are identified as re-vegetated. Re-vegetation of eroded peatlands has been identified as a major control on the sediment flux from gullied systems (Evans and Warburton 2005), therefore understanding of the long-term sediment dynamics of eroded systems is intimately linked to the nature of re-vegetation of bare ground.

Although much that has been written about peat erosion emphasizes the severity of the problem and the rapid degradation of moorland surfaces, there are also numerous references to natural re-vegetation of

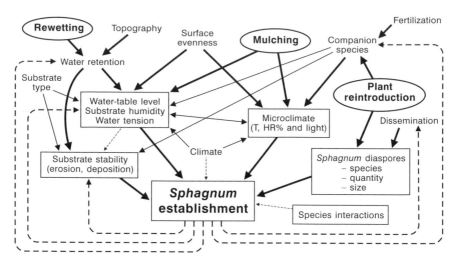

Figure 8.3   Key controls on the establishment of *Sphagnum* on peat surfaces (after Rochefort 2000)

bare peat sites produced either by erosion or by peat cutting. As early as 1930, White described re-colonization of Irish peat cuttings and Phillips (1954) described re-vegetation of eroding surfaces in the Southern Pennines. More recently several authors have noted natural re-vegetation in Irish bogs. Tomlinson (1981b) reports significant re-vegetation of eroded gullies in blanket peatland in Northern Ireland with *Eriophorum angustifolium* and *Sphagnum* forming the pioneer communities. Large and Hamilton (1991) and Cooper and McCann (1995) similarly refer to extensive 'healing' of eroded areas in central and Western Ireland, whilst Cooper et al. (2001) suggest that significant areas of blanket peat in Northern Ireland subject to historical peat cutting have recovered to vegetative compositions close to intact bog. A similar pattern is seen in the Republic of Ireland. In the Wicklow Mountains Bowler and Bradshaw (1985) identify formerly eroded blanket peat now covered in active peat-forming vegetation and Cooper and Loftus (1998) report that significant areas of the blanket bog system in the same area are eroded but re-vegetated to the extent that they are ecologically similar to the undissected areas they studied.

In the Pennines, Phillips (1954) suggests that re-vegetation was widespread on the slopes of the Bleaklow Plateau in the mid-twentieth century, particularly identifying the role of *Eriophorum angustifolium* in colonization of gully floors. Tallis (1969) similarly argues that *E. angustifolium* is important in the natural re-vegetation of gully lines. Evans et al. (2002) have demonstrated significant local re-vegetation at sites on Wessenden Head

Figure 8.4   Extensive cotton grass (Eriophorum angustifolium) re-vegetation of gully system at Shiny Brook, Wessenden Moor, South Pennines. (a) 1981. (b) Same site in 2002 (photos by permission, T. P. Burt and S. Clement)

Moor between the 1970s and the beginning of the twenty-first century (Figure 8.4), and Evans et al. (2005) identified widespread re-vegetation of formerly eroded surfaces across the Bleaklow and Kinderscout plateaux. Further north in the North Pennines and the Cheviot Hills the morphological evidence of severe erosion is considerable but the extent of bare peat is much less than in the Southern Pennines due to significant re-vegetation of eroded surfaces (Wishart and Warburton 2001; Evans and Warburton 2005).

Reports of natural re-vegetation of bare peat surfaces from elsewhere in the world, where natural peat erosion is uncommon, typically relate to areas of bare peat associated with peat mining, particularly in Scandinavia and Canada. Spontaneous natural re-vegetation of such sites has been reported by Salonen and Laaksonen (1994), Robert et al. (1999), and Lavoie et al. (2003). Although reports of the natural re-vegetation of eroded peatlands are well dispersed in space and time many of these reports are qualitative observations gleaned from publications where the re-vegetation is not the primary focus of the work. There is very little detailed work on the mechanisms of re-vegetation. This is an important omission since a proper understanding of natural re-vegetation processes should be a pre-requisite to work on the management and restoration of eroded moorlands. What follows is an attempt to draw together the limited work in this area and identify key controls on natural re-vegetation processes. This understanding is central to addressing the extent to which peat erosion may be partially self-limiting through processes of natural re-vegetation.

8.4   Controls and Mechanisms of Natural Re-Vegetation

The ultimate control on the establishment and spread of vegetation on any surface is the existence of local conditions consistent with the auteco-logical requirements of mire species. Considerable progress has been made on understanding the particular requirements of key species and some of this work is discussed above. However, a detailed consideration of autecological controls on re-vegetation is beyond the scope of this book. What is central to the discussion here is the interaction between geomor-phology and re-vegetation which is considered further below. The pro-cesses of natural re-vegetation are influenced by intrinsic and extrinsic controls. The extrinsic controls relate largely to the nature of the exter-nally imposed environment, whereas the intrinsic controls are a function of the natural development of eroded landscapes. Assessment of the long-term trajectory of eroded peatlands requires knowledge of both types of potential controls.

8.4.1   Extrinsic controls on re-vegetation

The extrinsic controls on re-vegetation can be divided into climatic and anthropogenic effects. What is clear from attempts to restore degraded mire is that maintenance of a high water-table is important to successful re-vegetation (Price and Whitehead 2001; Girard et al. 2002; Gorham and Rochefort 2003; Vasander et al. 2003). The role of climate is there-fore central in that the two dominant components of the peatland water

Table 8.1   Extrinsic controls on re-vegetation of eroded peat

Limiting factor	Mechanism	Reference
Grazing	Preferential grazing of young shoots of pioneer species especially in spring	Tallis and Yalden 1983; Grant and Armstrong 1993; Evans 1997; Evans 1998
Pollution	Acidification of surface peats. Seedling mortality due to direct pollution impact	Tallis 1964; Ferguson et al. 1978; Ferguson et al. 1987; Caporn et al. 1995; Richards et al. 1995; Anderson et al. 1997; Skeffington et al. 1997
Trampling	Low resistance of moorland vegetation to trampling, compaction of bare surfaces	Borcard and Matthey 1995; Anderson et al. 1997; MacGowan and Doyle 1997; MacGowan 2002
Climate	Lowered water-tables due to changes in water balance	Rouse 2000; Price and Whitehead 2001; Van Seters and Price 2001; Gorham and Rochefort 2003
	Seedling mortality due to frost or desiccation	Tallis and Yalden 1983; Anderson et al. 1997; Tallis 1997; Groeneveld and Rochefort 2005

balance, rainfall and evapotranspiration, are both climatically controlled (Chapter 2). Ombrotrophic mires exist because they maintain a positive water balance, and climatic conditions conducive to this state are therefore a pre-requisite to successful re-vegetation. Seasonal climatic patterns are also important because excessive summer desiccation or winter frost are also detrimental to the survival of seedlings (Groeneveld and Rochefort 2005).

Anthropogenic controls on re-vegetation may include grazing, pollution and, more locally, trampling (Table 8.1). As discussed in the previous section, work on the promotion of vegetation through management intervention has demonstrated that at some sites re-establishment may be limited by grazing so that exclusion of stock has a beneficial effect on rates of vegetation re-colonization of bare peat. Grazing pressure may be significant at relatively low stocking rates due to the tendency of stock to graze preferentially on young shoots of pioneer species. Tallis (1964b) suggested that the coincidence of soot layers and evidence of the decline of *Sphagnum* in peat cores from the Southern Pennines indicated a pollutant effect on the vegetation. Ferguson et al. (1978) demonstrated significant sensitivity of *Sphagnum* to sulphur pollutants and suggested that the near elimination of *Sphagnum* from the South Pennines was due to sulphur toxicity. Survey of re-vegetated erosion gullies in the South Pennines by Evans et al. (2005) showed *Sphagnum* as a common component of these communities consistent with recently reduced sulphate loadings (Skeffington et al. 1996). In contrast levels of nitrogen deposition remained high into the 1990s but are reducing. High nitrate deposition levels have been shown to depress *Sphagnum* growth but may encourage growth of grass species by improving the nutrient status of bare peat surfaces (Press et al. 1986; Lee et al. 1993; Hogg et al. 1995; Gunnarsson and Rydin 2000; Limpens et al. 2004). The effects of trampling are much more local, and at some sites may be severe. Moorland vegetation is particularly vulnerable to foot traffic (Pearce-Higgins and Yalden 1997). Trampling can lead to local removal of vegetation whilst compaction and continued foot traffic tend to limit re-vegetation of well-used pathways.

The extrinsic factors identified in Table 8.1 determine whether, for a given peatland, the external pressures on key pioneer species are limiting to re-vegetation. At the most basic level this may be measured by the extent to which spontaneous re-vegetation is observed in the landscape.

High pollutant levels during the industrial revolution are frequently cited as an important factor in the triggering of widespread erosion in the Peak District but as Tallis, Rhodes and others (Bowler and Bradshaw 1985; Tallis 1994, 1995; Rhodes and Stevenson 1997; Tallis 1997b and c) have demonstrated the erosion was underway well before the period of peak deposition, and occurred in regions such as Northwest Scotland and the North Pennines where pollution levels were much lower. What is distinctive about the Southern Pennines is not that there has been more erosion but that there has generally been less re-vegetation. The widespread evidence of natural re-vegetation of eroding peatlands suggests that a cycle of cut and fill, or erosion and re-vegetation, is a natural characteristic of blanket peatland surfaces. If this is the case it is possible that the severe and active gully erosion observed in the Southern Pennines represents a cumulation of natural erosion events due to reduced rates of re-vegetation and landscape recovery. In other words the effect of the pollution may not be to initiate erosion but rather to retard the recovery from erosion. It is arguable that sustainable artificial re-vegetation of peatlands is unlikely to succeed unless the external pressures are at sufficiently low levels to promote natural re-vegetation. The implication is that regional or national measures to reduce air pollution, and probably also to reduce grazing levels, are required prior to local attempts to promote re-vegetation of eroded peatlands.

Extrinsic factors typically act over a large spatial scale. The complex surface of a typical peatland is comprised of a wide range of microtopes with associated microclimate and micro-environmental conditions. For example, Grosvernier et al. (1995) note that the requirement of a high water-table for *Sphagnum* regeneration is locally modified by the creation of appropriately humid microclimates by pioneer species such as *Eriophorum vaginatum*. The assemblage of microtopes and hence microenvironments on the moor surface is strongly linked to the internal processes of growth, decay, erosion and re-deposition so that the local pattern of natural re-vegetation is largely a function of intrinsic controls.

8.4.2   Intrinsic controls on re-vegetation

Successful attempts to actively enhance vegetation of degraded peatlands require an understanding of local controls on re-vegetation. These might be termed intrinsic controls comprising the natural processes operating within gullied landscapes which tend to promote re-vegetation, and in particular affect the spatial patterns of natural re-vegetation. In an eroding mire, or one recovering from an erosion episode, the geomorphological microtopes dominate the mire surface so that the microscale hydrology and climatology of the mire surface is strongly correlated with its morphology (Chapter 7). The spatial distribution of environmental conditions suitable to the establishment of particular species is therefore strongly conditioned by geomorphological processes. This association can be examined further by considering the particular geomorphological contexts where natural re-vegetation is observed on mire surfaces.

Evans et al. (2005) hypothesized four principal modes of natural re-vegetation of eroded peatlands based on studies of mainly gully re-vegetation on the Kinderscout and Bleaklow plateaux of the Southern Pennines. These are:

1   Gully blockage – This involves partial blockage of gullies by bank failure which impedes drainage. This causes deposition of eroded peat and promotes establishment of *Eriophorum angustifolium*. This mechanism is similar to that proposed by Phillips (1954), who identified *Eriophorum angustifolium* as a key species colonizing mud pools and eroded gullies. Based on observations in the eroded peatlands of the Southern Pennines Phillips proposed a model in which dense growth of *Eriophorum* shoots on gully floors traps sediments creating zones of reduced gradient on the gully floor. Phillips rather confusingly describes these depositional forms as erosion steps and argues that the process acts to limit further erosion

of the gully. Stepped gully floor profiles associated with gully floor re-vegetation are common in the area today and Evans et al. (2005) have suggested an alternative interpretation (Figure 8.5a). Under this model the process is geomorphologically controlled. Collapse of oversteepened gully walls leads to impedance of flow along the gully floor. This promotes sedimentation behind the block and the *Eriophorum angustifolium* spreads vegetatively into the soft re-deposited peat. The longer-term development of these blocked sites is less clear. Because of the stepped profile created there is the potential that blocks and associated re-vegetated surfaces may be removed by headward migration of the steps due to fluvial erosion. However, the zone of active gully floor re-vegetation is also observed to propagate upstream as fresh sediment is trapped and the *Eriophorum* spreads vegetatively (Crowe unpublished data). The longer-term stability of gullies re-vegetated in this manner is most likely a function of the relative rates of step migration and upstream propagation of re-vegetation. Much more work is needed in this area.

2 Clump growth – This is characterized by gully re-vegetation by spread of slumped *Eriophorum vaginatum* clumps. These clumps are often observed uprooted in the base of steep eroding gullies and are mobile under conditions of high flow. A common pattern of re-vegetation involves a spread downstream of several such clumps over a distance of tens of metres (Figure 8.5b). It is hypothesized that initial clump formation is from seed or from vegetated failed blocks and that the re-vegetation is propagated downstream by the fluvial transport of mobile clumps.

3 Gully widening – The widening of gullies promotes local reduction of stream power allowing sediment deposition and re-vegetation. This mechanism was originally proposed by Wishart and Warburton (2001) in relation to re-vegetated gullies in the Cheviot Hills of Northern England. Once gullies have incised as far as the more resistant mineral sediments below the peat, rates of vertical incision are reduced. In these wider gullies gradient is reduced and the increase in gully width allows wandering or meandering of the flow. Reduced stream gradients reduce stream power and allow local redeposition of sediment, particularly on the inside of channel bends (Figure 8.5c). These zones appear to provide a suitable substrate and lower energy environment to allow re-vegetation, commonly by either *Eriophorum angustifolium* or *Eriophorum vaginatum*.

4 Peat-flat dead zones – On extensive low-angled areas of eroding peat ('peat flats') deposition of eroded peat in depressions and at the margins provides suitable substrate for the spread of *Eriophorum angusti-folium* (Figure 8.5d).

The particular species associations with these geomorphic locations might be expected to be a function of the location of this study in the eroded

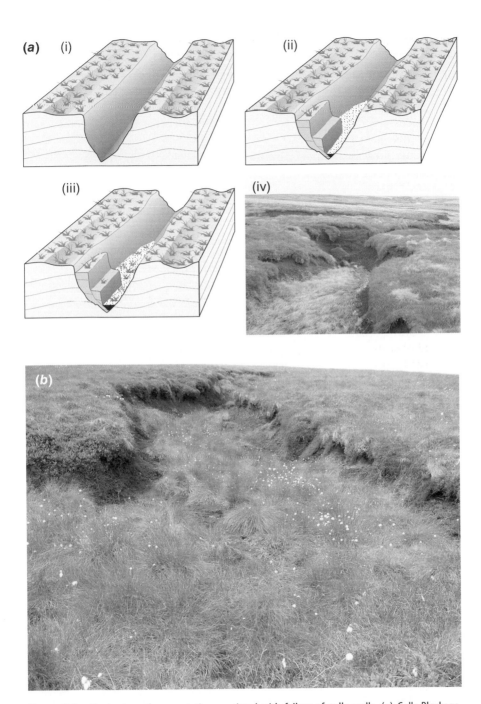

Figure 8.5 Mechanism of re-vegetation associated with failure of gully walls. (a) Gully Blockage. (i) Incision of deep gully. (ii) Failure of oversteepened gully wall partially blocks gully and impedes drainage. (iii) Sediment accumulation behind partial block provides suitable substrate for re-vegetation by *Eriophorum angustifolium*. (iv) Illustration of gully block and extensive *E. angustifoium* spread upstream from the Southern Pennines. (b) Clump Growth. Illustration of steep gully showing extensive re-vegetation by clumps of *Eriophorum vaginatum*.

Figure 8.5 *Continued* (c) Gully Widening. (i) Illustration of re-vegetation on the inside of bends in broad gullies. (ii) Example of gully re-vegetating by this mechanism on the Bleaklow Plateau in the Southern Pennines. (d) Peat Flat Dead Zones. Example of *Eriophorum angustifolium* colonization of the edges of a depositional zone on a peat flat

South Pennines. In other geographical locations different species will be important in the re-vegetation process but the general observation that re-vegetation is associated with particular geomorphological contexts should hold for any eroded peatland. Channel blocks, gully widening and the development of peat flats are natural consequences of phases of extended peat erosion. The importance of these geomorphological microtopes as loci of re-vegetation supports the view that once extrinsic limitations on re-vegetation are relaxed the natural trajectory of eroded peatlands is towards re-vegetation rather than continuing catastrophic erosion. One consequence of the identification of preferred morphological settings for re-vegetation is the potential to use easily available terrain data (LIDAR, stereo photography, published mapping) in planning management intervention to promote re-vegetation. This is simpler and cheaper than widespread direct measurement of the micro-environmental conditions which are the final determinant of the success of re-vegetation.

8.4.3 *Eriophorum* spp. as keystone species for re-vegetation of eroded peatlands

The central role of *Sphagnum* in the growth and functioning of ombrotrophic peatlands has long been recognized (Pearsall 1950), and Rochefort (2000) argues that *Sphagnum* is keystone species in the regeneration of mires. The experience of artificial re-vegetation suggests that, in addition to suitable water-table conditions and a source of *Sphagnum* diaspores, successful establishment requires a mulch to provide shade and increase humidity. Under conditions of natural mire re-vegetation it is commonly observed that initial colonization of bare peat surfaces is by species other than *Sphagnum*. The early pioneer species vary but may include *Eriophorum vaginatum*, *Eriophorum angustifolium*, *Eriophorum spissum*, *Polytrichum strictum* and *Polytrichum alpestre* (Maltby et al. 1990; Grosvernier et al. 1995; Richards et al. 1995; Robert et al. 1999; Lavoie et al. 2003; Evans et al. 2005). The role of *Eriophorum* spp. as a precursor to more diverse mire flora, and particularly in creating a suitable microenvironment for the establishment of *Sphagnum*, is most commonly reported (e.g. Tuittila et al. 2000; Lavoie et al. 2003). Sundberg and Rydin (2002) have identified *Eriophorum vaginatum* as a keystone species (Begon et al. 1996) for re-vegetation of bare peat. It is argued here that in fact this appellation should be extended more widely to include *Eriophorum angustifolium* which is a widespread early colonist of eroded and disturbed peatlands (Boudreau and Rochefort 1999; Richards et al. 1995; Cooper et al. 2001; Farrell and Doyle 2003). Canonical correspondence analysis of species associations in eroded and naturally re-vegetated gullies in the Southern Pennines has identified gradients from both *Eriophorum angustifolium* and *Eriophorum vaginatum* to more diverse flora including *Sphagnum* spp. (Evans et al. 2005). Recent work in the same area by Crowe (unpublished, Figure 8.6a and b) has demonstrated that these apparent trajectories observed in space are replicated chronologically in the stratigraphy of plant macrofossils from re-vegetated gullies.

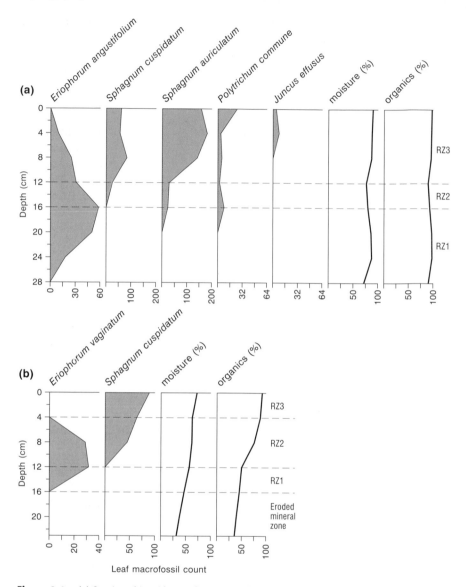

Figure 8.6 (a) Stratigraphic evidence of succession from *Eriophorum angustifolium* to a more diverse vegetation including *Sphagnum* on the floor of an eroded gully in the Southern Pennines (by permission of S. Crowe). (b) Stratigraphic evidence of succession from *Eriophorum vaginatum* to *Sphagnum* above the mineral floor of an eroded South Pennine gully. Horizontal axes are counts of leaf fragments

The ability of the *Eriophorum* spp. to rapidly colonize and stabilize bare peat surfaces as pioneer species is central to the argument that they are keystone species for re-vegetation of degraded mires. Evidence of succession to more diverse flora supports the view that increased stability and humidity under cotton-grass cover provides conditions suitable for other important moorland species, in particular *Sphagnum*, to establish (Rochefort 2000; Lavoie et al. 2003). There is however evidence that even in advance of succession to a stable diverse moorland community the *Eriophorum* cover plays a role in restoration of important mire functions. Tuitilla et al. (1999) have demonstrated that where water-tables are sufficiently high *Eriophorum angustifolium* mires in Finland are efficient carbon sinks. Similarly the *Eriophorum* species play a significant role in the retention of sediments on slopes reported by Evans and Warburton (2005) in re-vegetated peatlands in the North Pennines.

8.4.4 Re-vegetation dynamics and long-term patterns of erosion

Overall the extensive work on both natural and artificial re-vegetation of bare peat surfaces demonstrates a clear propensity for spontaneous re-vegetation when extrinsic controls permit. In eroded peatlands dominated by gully erosion the evolution of gully form over time tends to produce a morphology and substrate favourable to re-vegetation, whether through gully wall collapse and gully blockage or through gully widening and development of zones of sediment deposition. The implication is that under pristine conditions damaged mires show a natural trajectory towards re-vegetation and the re-initiation of peat formation. This is not a prescription for a *laissez-faire* approach to the management of eroding bogs, but it is a positive indication that, given the establishment of suitable boundary conditions (pollution climate, hydrological regime, grazing regime), well judged management interventions to promote re-vegetation will be working alongside the natural tendency for the mire surface to re-vegetate.

One area where further research is needed is into the relation between erosion and re-vegetation. Most of the research on peat erosion to date has focussed on identifying phases of breakdown of the surface vegetation layer and consequent erosion. The evidence for significant spontaneous re-vegetation of eroded peatlands implies that phases of widespread erosion could arise through the suppression of natural re-vegetation, and the accumulation of a mosaic of eroding surfaces, as well as through a general increase in the intensity of erosion processes. It is worth noting that the concept of cumulation of erosion impacts in the landscape is consistent with the conclusions drawn about Southern Pennine peat erosion by Tallis (Tallis 1997b, 1998), namely that the current severity and extent of erosion is the combined result of multiple impacts on the peatland in the last millennium. In fact many of the extrinsic factors which limit re-vegetation (such as pollution, grazing, climate change, fire, etc) are also those that have been identified as key factors triggering dis-integration of the vegetation cover and onset of erosion. An assessment of the relative importance of increased erosion and re-vegetation sup-pression will rest on identification of spatial patterns of the onset of erosion during the 'erosion phase' of the last millennium and in the pre-ceding period. This will require a major programme of dating erosion episodes and attempts to identify stratigraphic evidence of past phases of erosion.

8.5   Stratigraphic Evidence of Erosion and Re-Vegetation

The evidence of extensive natural re-vegetation of eroded peatlands pro-vides support for the hypothesis of periodic erosion and recovery of peat-land surfaces. Extensively eroded and re-vegetated peatlands in the North Pennines (Evans and Warburton 2005), the Cheviot Hills (Wishart and Warburton 2001), Galloway (Rhodes and Stevenson 1997), and in Ireland (Large and Hamilton 1991) provide a clear demonstration of a recent cycle of cut and fill. What is less clear is whether this is a historically unique phase associated with widespread acceleration of peat erosion in the last millennium (Higgitt et al. 2001) or whether there may have been earlier phases of erosion and recovery.

Thin (<10 mm) sand horizons are often observed within apparently stratigraphically intact exposures of peat in the UK. Bower (1962) sug-gested that these horizons must represent local deposition associated with upstream erosion of the sub-peat mineral surface and are therefore indica-tive of local phases of erosion. If this is the case the implication is that whilst the most severe peat erosion may be confined to the last millen-nium, local phases of erosion have occurred throughout the 3–8,000-year period of peat accumulation. These sand horizons are most commonly observed in gully fills, or in old flushes consistent with the view that they represent former erosion phases. Recent work however presents an alter-native interpretation for some of these features. Holden and Burt (2002a) suggest that pipes in deep peat often transport significant quantities of mineral sediment, presumably because at some point along the length of the pipe it is in contact with the mineral sub-peat surface. Splays of deposited mineral material are also often observed on the peat surface where pipes come to the surface. Given the frequency of piping in deep peat (Jones 2004; Holden 2005) these local pipe effluences provide an alternative explanation for some local sand layers within the peat stratigraphy. Similarly Caseldine et al. (2005) have identified silt layers in Irish bog sequences which are associated with mass movement on adjacent slopes with mineral soil cover.

If upslope erosion has occurred it should also be possible to detect stratigraphic evidence of organic sediment deposition at downslope sites since re-deposition of eroded peat within catchments is common particularly in actively re-vegetating gully systems (Chapter 4). This possibility has received relatively little attention, in part because of the difficulty of distinguishing re-deposited peat from *in situ* peat growth in the stratigraphic record. Work by Clement (2005) on recently re-vegetated gully systems in the Pennines and Cheviot Hills, UK, has demonstrated that the re-deposited peat is characterized by a combination of the following diagnostic properties, evaluated in relation to local undisturbed peat deposits:

1 less fibrous, more rounded peat particles due to sorting by wash processes;
2 few identifiable vegetation components due to greater oxidation under aerobic conditions and mechanical destruction;
3 low bulk-density values generally less than *in situ* peat due to lofted and openwork nature of the deposit (although greater mineral content offsets this to some degree);
4 low loss on ignition values (organic matter content) due to loss of organic rich material and incorporation of mineral fractions;
5 abrupt changes in macro fossils which correlate poorly with undisturbed reference sites;
6 higher sand and grit content often with distinct mineral lenses or pockets;
7 large peds or small blocks of peat in a finer peat matrix showing peat structures inconsistent with primary depositional fabrics, for example vertical stratigraphic horizons;
8 stratigraphic units which are lateral discontinuous or form lenses or pocket-type deposits.

It is important to recognize that, dependent on the local peat stratigraphy and geomorphic setting, these properties will vary and should only be used where knowledge of the undisturbed local peat stratigraphy is well known. There is increasing recognition by those working on bog stratigraphy that physical instability of peat surfaces is a possibility (Ashmore et al. 2000; Caseldine et al. 2005), and the development of appropriate techniques for recognition of eroded and re-vegetated sites (Clement

2005) suggests that this is a productive area for future work. At present on the basis of the very limited evidence available, there is no definitive evidence for extensive phases of peat erosion pre-dating the last millennium. However, there is the potential that with further work, peatland stratigraphy may, under favourable circumstances, reveal as much about geomorphological changes affecting the mire surface as it does about historic vegetation patterns or pollution history.

8.6   The Future of Blanket Peat Sediment Systems

This chapter has considered the long-term evolution of blanket peat sediment systems and the linkages between controls on mire surface vegetation and the geomorphological development of the peatland system. It is argued that eventual re-vegetation represents a natural state in the evolution of gully-eroded peatlands and that the extent of erosion observed in the landscape at any given time is a dynamic equilibrium between erosional processes and processes of re-vegetation. This provides the conceptual background for considering the potential future trajectory of peatland sediment systems, in response to predicted changes in environmental conditions. Climate change and changes in atmospheric pollution are two key factors identified as important to the long-term health of blanket mire (Peak District National Park Authority 2001).

8.7   Changes in Pollution

On a global scale the impact of pollution on mire surfaces is very localized. The erosion of the Southern Pennines has been associated with severe atmospheric pollution over the last 200 years (Tallis 1997). However, with progressive statutory limitation on emission of atmospheric pollutants, pollution in the UK has ameliorated. Skeffington et al. (1997) (Figure 8.7a) have demonstrated significant reductions in sulphate emissions and deposition over the second half of the twentieth century and Figure 8.7b shows local reductions in atmospheric nitrate and sulphate levels in the Peak District (UK) over the last 20 years. Skeffington et al. suggest that there has been a recent shift from sulphate to nitrate as the dominant phytotoxic atmospheric pollutant in the Pennines. Continuing decreases in nitrate concentration are likely to further reduce pollutant impacts. Together with reductions in grazing intensity due to local management interventions these changes have the potential to reduce the extrinsic limitations on spontaneous re-vegetation of eroded sites. Monitoring of the potential recovery of the South Pennines will be fundamental

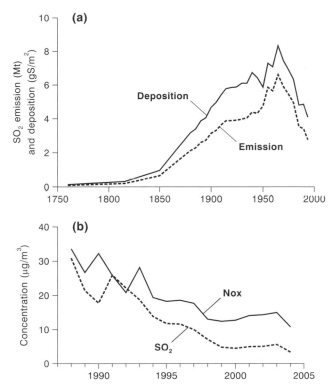

Figure 8.7   (a) Reconstructed UK sulphate emission and deposition spanning the industrial revolution (redrawn after Skeffington et al. 1997). (b) Declining sulphate and nitrogen concentrations in air measured at Ladybower reservoir in the Southern Pennines, 1985–2004. Data from the National Air Quality Archive

to understanding the interaction of erosion and re-vegetation should widespread erosion become a more frequent feature of marginal peatlands.

8.8   Climate Change Impacts

The major global risk to the physical integrity of upland peatlands is climate change. Ombrotrophic peatlands exist because climatic conditions are consistent with the maintenance of a positive water balance. Similarly the main natural factors promoting peat erosion are generally recognized as climatically controlled, namely, water, frost and wind (Philips et al. 1981). Consequently climate changes have the potential to

Table 8.2 Hydrological and erosional consequences of climate changes on upland peat in Britain. Climate change scenarios from Hulme et al. (2002)

Climatic change	Hydrological change	Erosional impact
Increased summer and autumn drought	Lower water-tables (greater acrotelm depth)	Peat shrinkage and desiccation Aeolian (dry blow) erosion
Increased summer and winter rainfall intensity	Increased peak storm runoff	Accelerated erosion of bare peat areas (rainsplash and wash) and increased channel erosion and gullying
Extended growing season	Greater evapotranspiration (minor)	Reduced erosion due to re-vegetation of bare peat areas and more mature vegetation blanket
Reduced frost frequency	Reduced impact of snowmelt events	Less frost-heave disturbance Less disruption to newly established vegetation

produce a range of erosion consequences. Scenarios of future climate change cannot be determined without uncertainty (UKCIP 2005), but general geomorphological responses to likely directions of change can be hypothesized (Table 8.2). The following discussion considers likely UK climate change scenarios from Hulme et al. (2002).

8.8.1   Increased summer drought

Summer drought has the potential to bring about large and to some extent irreversible changes to the peat landsystem (Burt and Gardiner 1984; Burt et al. 1990; Evans et al. 1999). Extreme drought will result in reduction of water-tables and desiccation of the surface peat. This leads to cracking and disaggregation of the crust into small peds which are vulnerable to wind erosion but also mechanical breakdown by trampling. Once desiccated, peat is very hard to re-wet and becomes detached from the peat blanket for an extended period when it can be easily transported. Gully systems are particularly at risk to erosion-desiccation processes because exposed faces dry quickly and, under the influence of gravity and wind, particles are rapidly removed and accumulate in gully bottoms. During subsequent flow events virtually all the sediment is mobilized and transported downstream. Another major factor governing peat loss and degradation during drought is the increased incidence of fire. The effects of fire on peatland systems are well documented (Imeson 1974; Mallik et al. 1984; Maltby et al. 1990). These can lead to greatly increased erosion through enhanced runoff and drainage dissection of the peat blanket. The recovery times from such events are very long and in some cases the peat is permanently damaged.

8.8.2   Increased winter rainfall intensity

Predictions of an increased proportion of rainfall falling in large storms may have severe short-term erosion impacts in the uplands. Upland peat landscapes are by their very nature high runoff areas and streams draining the peat blanket are often steep and flashy in nature. The impact of large storms is focussed dominantly on streams and gully systems. This can result in lateral stripping of the peat blanket and erosion at the margins of floodplains and in gully systems, often yielding large masses of peat blocks (Chapter 4). The other potentially more damaging effect is mass failures of the hillslopes adjacent to channels. Peat slides or bog bursts are a well documented phenomenon of upland peatlands (Chapter 5). These are major landscape-changing processes resulting in complete removal of the peat from large areas of the hillslope. Peat landslides deliver significant quantities of peat to upland stream systems resulting in channel modification and considerable ecological damage downstream. The remaining hillslopes are often left bare and can be susceptible to secondary erosion processes and gullying of the mineral substrate.

Chapter 5 identified an apparent increasing frequency of landslides recorded in British peatland environments. A severe storm may trigger multiple failures in a particular area (Carling 1986a; Warburton et al. 2003). Peat failures are important because of their short-term severe impact on stream ecology and permanent long-term degradation of the peat carbon resource. Although shallow failures in mineral soils are often quick to recover, peat failures represent a permanent loss of the surface soil. The recent rise in failure frequency is associated with a small number of significant landslide clusters occurring on a local scale and often triggered by single storms. This suggests that although the frequency of individual triggering events is increasing slowly, the impacts of such events on the landscape may be proportionately greater.

8.8.3   Changes in the growing season and re-vegetation

As climate changes the vegetation will respond in several ways. Of greatest significance for peatland erosion are probably the extension of the growing season, the moisture balance of the peat soils and changes in species composition. Warming temperatures will generally result in an extension of the growing season and a reduction of the frequency of frosts (Holden 2001; Holden and Adamson 2002). An extended growing season allows vegetation to develop for longer and reduced frost frequency will increase seedling survival. The net effect is likely to be the progressive re-vegetation of bare peat soils. However, this can only be achieved alongside grazing management (see 8.4.1 and Shaw et al. 1996).

8.8.4   Reduced frost frequency

The importance of frost weathering in sediment production has been identified in Chapter 3. Reduced sediment supply may act to counter increased erosion due to enhanced rainsplash activity. Reduced frost action together with warmer summers will tend to shift the seasonal balance of sediment production towards the summer. The studies by Francis (1990) on relatively low elevation peatlands in Wales during a drought period which emphasize autumn export of sediment generated through summer desiccation provide a potential analogue for this scenario.

8.8.5   Overall response of the peatland system

Upland peatlands have formed due to a unique set of conditions brought about by a net surplus in the water balance equation. Climate change will alter this balance. If we conceive of sensitive (Bragg and Tallis 2001) upland peat landsystems as a dynamic equilibrium of erosional processes and re-vegetation processes then the net impacts of climate change will be dependent on the net response of the geomorphological and ecological systems. Lower water-tables will reduce resistance to erosion and increased storminess will accelerate erosion, but longer growing seasons may favour re-vegetation. The ability of vegetation to stabilize the peat surface may however be compromised by increased fire frequency. The overall effect is likely to be a more dynamic bog surface characterized by a mosaic of eroding and recovering patches. The significance of these changes will have important implications for peat loss from the uplands and the overall carbon balance (see Chapter 9) (Worrall et al. 2003).

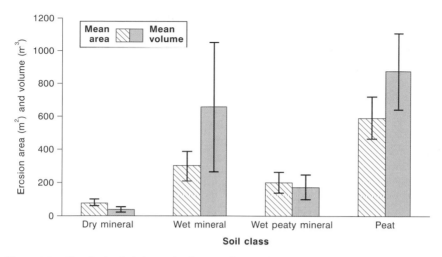

Figure 8.8 Magnitude of obvious upland erosion features by volume for various soil classes (after McHugh et al. 2002)

8.9 Relative Importance of Peat Erosion in Wider Upland Sediment Budgets

Figure 8.1 implies that there have been significant changes in the nature of peatland sediment budgets throughout the Holocene, and that there is the potential for dramatic changes in the future, particularly if the impacts of predicted climate change are lower water-tables and enhanced erosion. It is important to consider therefore the role that peatland sediment budgets play in wider upland sediment systems. In considering this issue the case of the British upland peats can be taken as a worst case scenario. A third of Britain is classified as uplands and about 23 per cent of this landscape is covered by blanket mire (Tallis et al. 1997; Fielding and Howarth 1999), much of which is heavily eroded. Reported sediment yields from eroding peatlands (range 50–265 t km^{-2} a^{-1} see Table 4.1) cover a similar range to the overall pattern for UK uplands reported by Walling and Webb (1987). This comparison to some extent masks the extent of the topographic impact of peat erosion on the upland landscape, since as Labadz (1991) noted the low density of peat (typically 0.1 t m^{-3} compared with circa 1.5 t m^{-3} for mineral soil) means the volumetric impact of peat loss is greater (Warburton et al. 2003).

McHugh et al. (2002) report that of 'obvious' erosion features recorded in a survey of the uplands of England and Wales almost 50 per cent by volume occurred in peatlands (Figure 8.8). In specific locally eroding

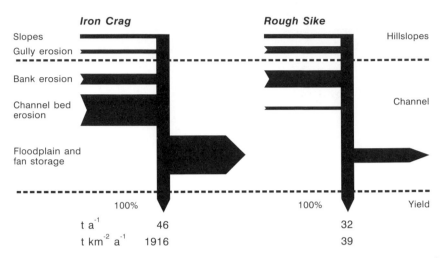

Figure 8.9   Outline sediment budgets for small catchments in northern England dominated by mineral (Iron Crag) and organic (Rough Sike) sediment respectively (after Warburton et al. 2002)

sites sediment yields from mineral ground can significantly exceed the catchment average values reported by Walling and Webb (1987). For example, Johnson and Warburton (2002) report sediment yields of $1,900 \, t \, km^{-2} a^{-1}$ from a steep gully-head catchment in the English Lake District. These data make an interesting contrast with the eroding peat-lands of Rough Sike (Warburton et al. 2003; Evans and Warburton 2005; Figure 8.9). The total sediment yield of the mineral catchment is a factor of seven greater than that of the peatland catchment, but given the density differences this implies greater topographic expression of erosion at the peatland site, even in comparison with a heavily eroded mineral sediment site. Taken together with the greater spatial extent of peat gullying com-pared with local debris-flow activity this further emphasizes the import-ance of upland peat degradation in total upland erosion identified by McHugh et al. (2002). Comparison of the two sediment budgets shows that in both the mineral- and the organic-dominated catchments hillslope erosion, including gullying, is a relatively small part of the total sediment yield. Sediment production by in-channel processes is the dominant com-ponent of the budget. Harvey (1994) has argued that the key control on fluvial sediment yield from areas of gullied mineral soil in the uplands of the Howgill Fells, northwest England, is the degree of linkage between the slopes and channel. These findings support the observation of the general importance of slope-channel linkages for sediment transfer in upland Britain. This is a pattern that has been proven to extend to peat-covered catchments as well (Evans and Warburton 2005). Harvey (1994) highlights the importance of re-vegetation of eroding gullies as a control on slope-channel linkage which is in line with the observations of the role of gully floor re-vegetation at Rough Sike described in Chapter 4, and emphasizes the importance of understanding the processes of re-vegetation of eroded peatlands, for the understanding of long-term sediment dynamics in upland landscapes more generally.

8.10   Conclusions

The last millennium has seen dramatic changes in sediment delivery from upland peatlands, particularly in the UK and Ireland. The most severe ongoing erosion occurs in the climatically marginal and severely anthropogenically impacted peatlands of the Southern Pennines. The long-term evolution of eroded peatlands is closely linked to the regional capacity for re-vegetation of eroded sites.

Tallis and Yalden (1983) suggest that the results of trials of artificial re-vegetation in the Southern Pennines are grounds for rejecting a 'fatalist' view of peat erosion. They argue that suitable management can promote re-vegetation of eroded peat surfaces.

It has been argued here that under suitable conditions (particularly under suitable climate and limited pollution impacts) spontaneous re-vegetation can also play a significant role in regenerating upland peatlands. The available evidence on peatland re-vegetation supports a hypothesis that spontaneous re-vegetation is a normal system response to erosion as long as external conditions are not limiting. The most severe manifestations of the fatalist view, or what was termed here the catastrophic model of peat erosion, are not supported by the small but growing body of work on peatland re-vegetation. In fact increased understanding of processes of spontaneous re-vegetation indicates that the suppression of re-vegetation processes may play a significant role in generating the extensive eroded landscapes of the past millennium alongside increases in the rate of erosion. The implication is that studies of peat erosion should be as much concerned with understanding of the dynamics of eroded and eroding systems as with identifying the threshold conditions for the onset of erosion. The former have received considerable attention in the literature (Chapter 1), but the latter is an area of investigation which urgently requires further work. The urgency for this research stems from the immediate requirement to assess and potentially respond to the effects of predicted climate change on the sediment budget of upland mires.

The worst-case scenario is that the severe impacts on climatically marginal peatlands in the UK might be replicated more widely in the upland peatlands of the Northern Hemisphere due to the effects of climate change. Ombrotrophic mires are climatically determined landform assemblages. Long-term shifts to negative water balances will eliminate the mires through desiccation and physical and chemical degradation. Furthermore, many northern peatlands are underlain by permafrost and continued warming is leading to destabilization of the upper peat mantle (Beilman and Robinson 2003; Huscroft et al. 2004). In deep mires where natural peat erosion processes are already highly effective, thawing permafrost has the potential to further enhance the efficiency of these mechanisms (Carey and Woo 2002). However, even more limited changes in the seasonality of the water balance have the potential to promote physical instability of the mire surface. Widespread erosion may require intervention to mitigate both on- and off-site negative effects of erosion. A clear understanding of interactions between erosion and re-vegetation is a prerequisite to appropriate management interventions in upland peatlands. The effects of erosion and potential management strategies are considered in the next chapter.

Chapter Nine

Implications and Conclusions

9.1 Implications of Widespread Peat Erosion

The consequences of peat erosion have been well summarized by Philips (1981) and include degradation of moorland ecosystems, loss of reservoir capacity, problems of colouration in water supply, loss of grazing land/ grouse habitat, and impacts on recreational use. In this chapter rather than revisit these issues we focus on three areas of recent concern where the characteristics of the upland peatland erosional system, explored in this book, have significant implications. These are: peat erosion and carbon budgets; release of stored contaminants from eroding peatlands; and restoration of eroded upland mire systems.

9.2 Upland Peatland Erosion and Carbon Budgets

Soils constitute by far the largest terrestrial carbon store and northern peatlands contain 20–30% of world soil carbon (Gorham 1991). Quantifying the carbon flux from these systems is therefore vital to understanding global carbon cycling. The high carbon content of peat coupled with its high erodibility can result in rapid loss of carbon from the terrestrial ecosystems (Milne and Brown 1997). Lal (2003) has recently argued that although erosion-induced carbon emissions are of global significance they are still poorly understood and quantified, and that it is important to examine carbon loss at all stages of the soil erosion process.

Hope et al. (1997a) identify fluvial export of carbon in both dissolved and particulate forms as a key component linking terrestrial peatland carbon stores to the ocean carbon sink (Hope et al. 1997a). Relatively few studies have considered both the particulate and the dissolved forms of

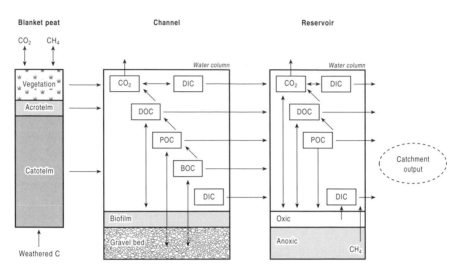

Figure 9.1   Simplified diagram showing the carbon pathways between the blanket peat C store, fluvial system and reservoir storage. BOC refers to block organic carbon which is a new term to describe peat blocks which are commonly transported in upland fluvial systems but not usually included in C balance estimates. DIC is dissolved inorganic carbon (Evans and Warburton 2001)

carbon flux together, but observations in lower reaches of several river systems suggest DOC is greater than POC, while towards the headwaters the relationship is reversed (Hope et al. 1997b). Establishing the significance of POC in upland catchments is therefore fundamental to understanding the carbon flux to river systems. Physical erosion is the principal source of POC in degraded peatlands (Figure 9.1). The release of carbon from upland peat depends on the nature of the peat store and the hydrological and geomorphological processes operating within the catchment (Aitkenhead et al. 1999). Understanding these processes is therefore central to establishing representative carbon budgets.

Unfortunately, there are considerable uncertainties in the current understanding of upland peatland carbon budgets. Four key areas of uncertainty exist:

1   Mechanisms of POC transport and transformation in fluvial systems are poorly understood, and POC is often disregarded as a significant component of the carbon budget.
2   Most of the work on POC and DOC flux has been undertaken in relatively intact peatlands, therefore the potential role of POC in eroding peatlands has not been fully explored.

3   In many carbon flux studies, particularly in the UK, sampling fre-
    quency has been inadequate to demonstrate the POC dynamics, or
    potentially, given the episodic nature of sediment transport in peat-
    land streams, even to capture representative values (Hope et al. 1994;
    Worrall et al. 2003; Armstrong 2005).
4   Approaches to estimation of sediment carbon contents vary. Normally
    POC is indirectly estimated from the suspended sediment concentra-
    tion or from loss on ignition values from filter paper residues (Hope
    et al. 1997b).

Despite these uncertainties sediment budget models can be used to determine the main sources and fluxes of sediment in upland catchments
by identifying and quantifying the main sediment storage and transport
processes operating in a drainage basin (Evans and Warburton 2005).
These models therefore provide an appropriate tool for understanding
how peat is eroded and the mechanisms by which sediment (POC)
enters streams (Figure 9.1). Coupling of sediment budget models with
carbon storage assessments (the carbon content of sediment stores)
provides a method for estimation of fluvial carbon flux from a given
catchment.

9.2.1   Case study example: the Rough Sike carbon budget

Here we present a case study, based on recent work by Evans and Warburton (2005), of an eroding peatland sediment budget in the North
Pennines, UK. We demonstrate how the relative importance of POC
contributions to the overall carbon budget can be evaluated. We also use
this example to explore how the carbon flux is estimated from measurements of fluvial suspended sediment transport, the importance of characterizing different sediment sources when estimating POC and linkages
between POC source and sink in the upland fluvial system.

   The case study explores the carbon budget implications of the Rough
Sike sediment budget described in Chapter 4. Rough Sike is a small catchment ($0.83\,km^2$) of the upper River Tees (northern England). The catchment is blanket peat covered and 17 per cent of the cover is eroded
(Garnett and Adamson 1997). The sediment budget is based on measurements of erosion and deposition at key sites in the Rough Sike catchment,
and on sediment yields derived from sediment rating curves applied to
four years of discharge data from the site (Evans and Warburton 2005).
A summary sediment budget for Rough Sike is presented in Figure 9.2.
The fluvial suspended sediment flux under contemporary conditions is

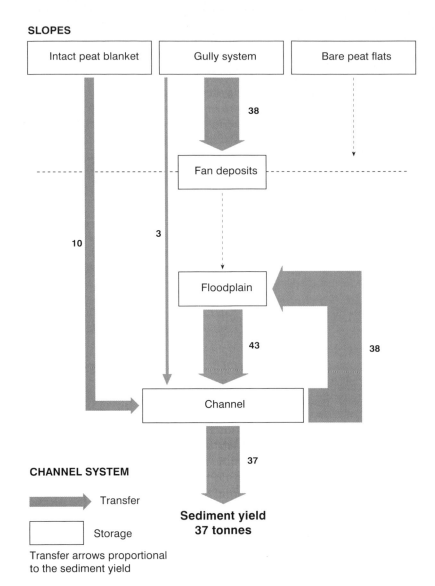

Figure 9.2 Sediment budget model for the Rough Sike catchment, Moor House, North Pennines

Table 9.1 Estimates of the present C balance for the blanket peat covered Moor House Catchment. Three models are presented showing uncertainties in the estimates of the POC flux (based on the work of Garnett [1998] and Garnett et al. [2000, 2001] and Evans and Warburton [2005]). The table shows the results of three mass balance models for the Moor House catchment. One caveat to these models is that net primary productivity is held constant. Under conditions of decreased pollution and grazing and a warming climate this may have increased. This area needs further research

C flux component	Rate $(gCm^{-2}yr^{-1})$		
	Model A	Model B	Model C
Net Primary Production minus decomposition	+27	+27	+27
SCP deposition	+0.03	+0.03	+0.03
Dissolved Organic Carbon	−10 to −20	−10 to −20	−10 to −20
Particulate Organic Carbon	−56	−19	−12
Net carbon balance	−39 to −49	−2 to −12	5 to −5

Notes:
Model A – Based on Garnett (1998) using an estimate of POC flux based on Crisp (1966).
Model B – Based on Garnett (1998) but using an estimate of POC flux based on Evans and Warburton (2005).
Model C – Based on Garnett (1998) but using an estimate of POC flux based on Evans and Warburton (2005) with a lower estimate (30%) for the carbon content of the particulate load.

controlled to a large degree by channel processes. Although gully erosion rates are high in the context of UK upland environments, poor connectivity between the slopes and the channels minimizes the role of hillslope processes in generating catchment sediment yield. Notably bare areas of peat flat, which are conventionally thought as the areas of maximum potential erosion, contribute little to the overall sediment flux. Comparison with historical field data (Crisp 1966) suggests that current sediment transport rates of $44\,tkm^{-2}a^{-1}$ represent a 60 per cent reduction in fluvial suspended sediment yield from Rough Sike over the last 40 years, which is consistent with reduced slope-channel linkage due to gully mouth revegetation (Chapters 4 and 8).

These changes have produced a transition in the carbon balance of the blanket peat mass from carbon source to carbon neutral. Table 9.1, based on Garnett (1998), Garnett et al. (2000, 2001) and the sediment budget from Figure 9.2, shows the results of three carbon-balance models for the Moor House catchment in northern England. In the absence of more recent data Garnett (1998) used sediment yield estimates based on the work of Crisp (1966) to estimate POC. Under this scenario, high rates of sediment flux result in a large net export of carbon from the catchment (Model A). However, using revised sediment budget data the net carbon export is greatly reduced (Model B, based on Evans and Warburton [2005]).

The sediment budget suggests that the source sediment will have an important bearing on the nature of the material supplied to the channel system, particularly with regard to the amount of organic material transferred. Sampling of the main sediment sources within the Rough Sike catchment show marked contrasts in patterns of organic matter storage: peat banks (organic content 93–99%), floodplain sediments (14–26% organic content) and channel sediment (4–7% organic content). Tapping of different sediment sources within particular storm events will produce differing organic matter fluxes.

Given the fact that most of the sediment is derived from channel and floodplain sources (Figure 9.2), which have lower carbon contents (~30%), the model can be further refined producing an estimated balanced budget (Model C). The apparent shift from the catchment as a carbon source to carbon neutral demonstrates the fundamental importance of understanding sediment budget models, and correct estimation of POC when calculating the carbon balance for upland catchments. Comparison with estimates of DOC flux (Garnett et al. 2000) indicate that even under contemporary conditions, where the main carbon store is disconnected from the fluvial system, POC is of equivalent magnitude to DOC. Under better connected and more actively eroding historic conditions the POC flux is three times contemporary DOC flux. Comparison of these POC data with recent models of the carbon budget at Moor House (Worrall et al. 2003) demonstrates that fixation of CO_2 by vegetation and the POC flux are the most important values in the contemporary peatland carbon balance.

Establishing the sources of POC is an important part of understanding the upland carbon budget; however, understanding the fate of POC when it enters the upland stream channel is also important (Figure 9.1). Our observations suggest that storage of POC in the river is negligible and sampling of river gravels generally shows low values of particulate organic material (<3%), although some deposition of peat occurs overbank in the form of large peat blocks (Evans and Warburton 2001). Much of the POC is likely to be transported downstream until it reaches the first major upland sink, which in many instances is a reservoir. Smith et al. (2001) have recently determined that the primary fate of eroded soil carbon in the conterminous United States is in impoundments. POC losses during fluvial transport have also been documented. For example, Dawson et al. (2004) note a large downstream decrease (–55%) in POC flux from small headwater streams in northeast Scotland. These changes are thought to be associated with the in-stream processing of POC in the water column rather than deposition within the channel (Figure 9.1). Both the mechanisms of POC delivery to upland streams and in-stream transformations are important areas for future research.

9.3   Release of Stored Contaminants from Eroding Peatlands

Ombrotrophic mires receive material inputs only from the atmosphere. They are therefore highly sensitive to the nature of atmospheric deposition and, because of the efficiency of organic materials in absorbing pollutant ions, upland peatlands can represent significant stores of pollutants. The main pollutant types are: sulphate and nitrate deposition from vehicle exhausts and from the burning of fossil fuels (Campbell and Lee 1996); metal deposition from fossil fuel, particularly coal burning, and from vehicle emissions (Weiss et al. 1999); and ammonia deposition associated particularly with livestock production (Apsimon et al. 1987; Sutton et al. 2003). Heavy metals have been particularly well studied in this context because of their toxicity at low concentrations and the extent to which they are tightly bound in organic soils. There is a large and diverse literature on peat bogs as archives of heavy metal pollution (e.g. Shotyk et al. 1997). Because of the relative immobility of these pollutants in the peatland environment the stratigraphic profile of pollutant concentrations preserves the historic patterns of metal deposition. This property has been widely exploited to reconstruct records of past industrial pollution (e.g. Lee and Tallis 1973, Livett et al. 1979; Jones and Hao 1993; Shotyk et al. 1998; Vile et al. 2000; Shotyk, 2002; Weiss et al. 2002).

Total metal loadings of bog surfaces can reach very high levels. Rothwell et al. (2006) tabulate peak lead concentrations recorded from a range of bogs around the world (Table 9.2). High concentrations occur in a range of locations but many of the highest values relate to the eroding peat bogs of northern England. Typically these studies show peaks in heavy metal concentrations in the upper 5–40 centimetres of the peat profile, reflecting peak emissions during the period of the industrial revolution. In most early industrializing areas contemporary rates of atmospheric pollutant deposition have been reduced by legislation (Mylona 1993), which is reflected in reduced concentrations of pollutants in the surface layers of peat profiles. However the total amount of pollutant stored in these upper peat layers can be very high and there is a significant risk that physical or chemical instability of the peat mass will remobilize these materials into the aquatic environment.

Despite the relative immobility of these pollutants release of dissolved metals in significant concentrations does still occur. Tipping et al. (2003)

Table 9.2   Maximum lead concentrations recorded in global peat bog sediments (after Rothwell et al. 2006)

Location	Maximum Pb concentration (mg kg⁻¹)	Author(s)
Fenno-Scandia tundra – forest-tundra zone, Russia	1650	Zhulidov et al. 1997
Alport Moor, Peak District, England	1647	Rothwell 2006
Gola di Lago, Switzerland	1528	Shotyk 2002
Ringinglow Bog, Peak District, England	1230	Jones & Hao 1993
Fairsnape Fell, Forest of Bowland, England	845	Mackay & Tallis 1996
Tinsley Park Bog, Lower Don Valley, Sheffield	827	Gilbertson et al. 1997
Grassington Moor, North Yorkshire, England	800	Livett et al. 1979
Ringinglow Bog, Peak District, England	700	Markert & Thornton 1990
Snake Pass, Peak District, England	570	Lee & Tallis 1973
Ringinglow Bog, Peak District, England	548	Jones 1987
Kola Peninsula, Russian Arctic	510	Zhulidov et al. 1997
Thorter Hill, Grampian Highlands, Scotland	489	Farmer et al. 2005
Boží Dar, Czech Republic	479	Vile et al. 2000
Lochnagar, Scotland	400	Yang et al. 2001
Ystwyth Valley, Cardiganshire, Wales	350	Mighall et al. 2002
Langmoos Bog, Mondsee, Austria	230	Holynska et al. 2002
Hajavalta, Southwest Finland	204	Nieminen et al. 2002
Rouyn-Noranda, Quebec, Canada	155	Kettles & Bonham-Carter 2002
Myrarnar, Faroe Islands	111	Shotyk et al. 2005
Ovejuyo Valley, Andean Royal Belt, Bolivia	23	Espi et al. 1997

have modeled the effects of drought on the release of dissolved metals from blanket peatlands in Scotland and northern England. They demonstrate that oxidation of sulphur (itself a stored pollutant in the upper layers of many bogs), and a reduction of pH under conditions of water-table depression, can lead to an order of magnitude increase in heavy metal flux. These results are particularly significant for eroding peatlands where the degree of water-table depression associated with local drawdown by gullies significantly exceeds that of even the most severe droughts (see Figure 2.12). The onset of gully erosion and local drainage therefore has the potential to significantly increase the rate of dissolved pollutant export. The risk associated with toxic heavy metals in the aquatic system is closely related to the speciation of the metals. A large part of the metals are transported bound to dissolved organic matter and consequently have relatively low bio-availability. However, Winch et al. (2002) have shown that UVB exposure has the potential to increase the proportion of free metals due to degradation of DOC. Increased free metal concentrations are also associated with lower pH (Gerhardt 1993), which may result from water-table drawdown.

In addition to the link between gullied peat and lower water-tables there is a direct geomorphological control of pollutant flux to stream systems. In eroding catchments the physical processes of erosion control material flux from the contaminated peat store to the stream system. High rates of organic sediment flux in eroding peatland systems (Table 4.1) have the potential to transfer significant amounts of stored pollutants to the aquatic system in bound particulate form. Rothwell et al. (2005) have demonstrated average sediment-associated lead concentrations of circa 100 mg kg^{-1} and peak values of 300 mg kg^{-1} in storm-water samples from an eroding peat catchment in the Southern Pennines (Figure 9.3). Use of magnetic tracing techniques proved that the lead was associated with erosion of a contaminated layer in the upper 20 centimetres of the peat profile, where lead concentrations exceeded 1,000 mg kg^{-1}. Peak in-stream lead concentrations were associated with greater erosion of this layer. In areas of gully erosion such as that studied by Rothwell et al. (2006), in-stream sediment concentrations are consistent with conservative mixing of contaminated upper peats with 'clean' peats from lower in the profile exposed on the gully walls. In areas where sheet erosion of the surface layer is occurring, much higher contaminant concentrations might be expected. Many of these areas of sheet erosion occur as a result of uncontrolled wildfires removing vegetation cover and subsequently enhancing rates of erosion The range of reported surface retreat rates for areas of sheet erosion (Table 3.1) is 18–41 mm a^{-1}. If the legacy of contamination from the industrial revolution is associated with a surface layer representing peat growth in the last 2–300 years then typically this may be eroded completely in 5–10 years resulting in very high contaminant loadings to the aquatic system. Rothwell et al. (2005) note that average values of lead contamination in the Southern Pennines exceed Canadian guideline values for probable effects on the aquatic ecosystem. The potential for ecosystem exposures far in excess of these average values in recently burnt

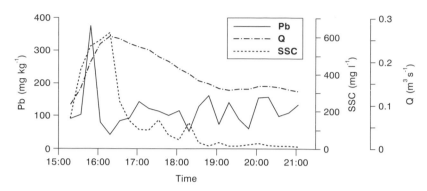

Figure 9.3   Export of sediment-associated lead during a winter storm event on Upper North Grain, a small peatland stream draining heavily contaminated blanket peat in the Southern Pennines (after Rothwell et al. 2005). SSC, Suspended Sediment Concentration. Q. stream discharge

moorland areas is significant. The ultimate fate of sediment-associated pollutants is controlled by the nature of fluvial sediment transport and potential in-stream transformations of the sediment (Figure 9.1). Organic sediment derived from peat erosion is deposited downstream on flood-plains (Evans and Warburton 2001, 2005) and in lakes and reservoirs (Holliday 2003; Labadz et al. 1991). Production and transformation of DOC in these locations is potentially a further important source of organically-bound dissolved metal pollution in the aquatic environment (Koelmans and Prevo 2003; Rothwell et al. 2006).

Erosion of blanket peat has the potential to release stored contaminants into the aquatic system, either indirectly through drawdown of water-table local to gullies or directly through sediment-associated contaminants. The relative importance of these mechanisms is a function of the pattern of peat erosion. Where extensive sheet erosion of surficial sediment occurs, for example on fire scars, contaminant release will be predominantly associated with sediment flux. Where the principal erosion

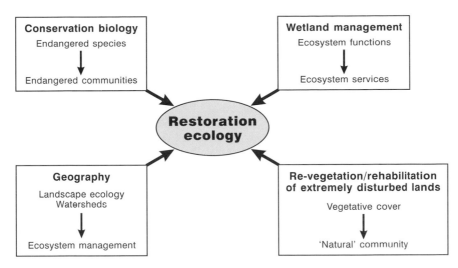

Figure 9.4   Four main strands of thought in restoration ecology. The four traditions differ in the components of the landscape system they prioritize for restoration (after Ehrenfield 2000)

type is gullying, release of contaminants in soluble form, bound to DOC, will be relatively more important.

9.4   Restoration of Eroded Upland Peatlands

9.4.1   Frameworks for restoration

Given the negative consequences of peat erosion, restoration of eroded mires is an important land management imperative (Wheeler et al. 1995). Part of the difficulty of mire restoration arises from the variable motivations for restoration. Ehrenfeld (2000) identified four main strands of thinking in restoration ecology focusing on restoration of ecosystem function, vegetation cover and watersheds, and support for rare species (Figure 9.4). It is arguable that because of the range of negative consequences of erosion each of these is relevant to the restoration of upland mires.

In the following sections the implications of an understanding of peatland geomorphology for approaches to peatland restoration have explored. The discussion is firmly rooted in what Ehrenfeld (2000) calls the geographical tradition of watershed, or landscape management, and reflects a geomorphological concern with landscape rather than with individual species.

9.4.2   Approaches to restoration

Attempts to restore upland peatlands can be divided into three main groups:

1   restoration of mined peatlands;
2   restoration of drained peatlands;
3   restoration of eroded peatlands.

Thorough reviews of mine restoration and of drainage restoration are provided by Charman (2002) and Holden et al. (2004) respectively.

One of the dominant approaches in restoration ecology has been to restore the physico-chemical system and to promote natural succession to a restored vegetation cover (Suding et al. 2004). In mined and drained systems the water-table has been widely identified as the key physico-chemical parameter for restoration. High water-tables are the condition which initially drives peat accumulation and to which mire vegetation is adapted. Therefore, restoration of water-tables should have the effect of promoting re-vegetation and restoring many important wetland functions. Van Seters and Price (2002) suggest that whilst establishment of vegetation cover occurs relatively rapidly on abandoned peat workings, re-establishment of hydrological function takes well in excess of 100 years in the absence of management intervention. Restoration strategies therefore frequently involve blockage of drainage ditches, creation of bunds or even contouring of the site (Wheeler et. al 1995; Stoneman and Brooks 1997; Price et al. 2002). Re-vegetation is promoted by seeding or planting with appropriate species often in with some type of mulch or nursery crop to assure appropriate microclimatic conditions at the surface (Price et al. 1998).

In contrast to mined peatlands there has been less work on the restoration of eroded peatlands. In common with drained systems there has been much inventive practical conservation undertaken by land managers (Holden et al. 2004), but techniques of restoration are not standardized and there has been relatively little post-restoration monitoring. The largest single body of work on restoration of eroded peatlands comes from the heavily degraded blanket peats of the Southern Pennines. Three reports on moorland erosion commissioned by the Peak District National Park (Philips et al. 1981; Tallis and Yalden 1983; Anderson et al. 1997) systematically examined the nature of erosion and restoration trials in the area. Much of this work focused on the extensive bare peat flats created by uncontrolled wildfire. Techniques including seeding, fertilization, and mulching have been developed to successfully establish *Calluna vulgaris* or *Deschampsia flexuosa* (Anderson et al. 1997). The techniques of drain blocking (Stoneman and Brooks 1997) have also been extended to these eroded areas through trials of gully blocking. Low dams built across eroding gullies impede drainage and provide stability with the aim of raising the water-table and promoting re-vegetation of gullied systems (Evans et al. 2005).

The findings of the case study outlined in Section 9.2.1 have practical implications relating to upland erosion control since they strongly justify gully-blocking-based approaches as a means of managing moorland erosion and carbon flux. The blanket peats of the Rough Sike catchment are extensively eroded but, in contrast to the peat moorlands of the Southern Pennines, they are notably more re-vegetated. Rough Sike lies towards one end of a spectrum ranging from eroded but re-vegetated peatlands to sites of active erosion. Natural re-vegetation of eroding areas of peat has led to the trapping of particulate material on the slopes and reduced transfer to streams. The efficacy of this process in reducing the particulate carbon flux from the system suggests a mechanism for minimizing carbon losses from eroding peatlands using relatively simple field management techniques such as gully blocking.

9.4.3 Implications of the landsystems model and sediment budget work for restoration

As noted above, one of the key questions facing any restoration project is what is the desired restoration end point? This question is particularly pertinent to eroding upland peatlands in the UK because these are semi-natural systems, many of which exist because of early anthropogenic impacts on the uplands and are maintained in their present form by a regime of burning and grazing. As a consequence periodic erosion and stabilization may in fact be part of the natural development of these mires. There is no doubt that in the most severely eroded mires a variety of anthropogenic impacts have combined to shift the balance between mire stability and erosion. Management strategies are therefore required to mitigate the negative effects of excess erosion but a restoration strategy that aims to return the entire mire surface to a stable uneroded condition may be unrealistic, both practically and in terms of the 'natural' state of the mire. Anderson et al. (1997) recognized this and argue that for the Southern Pennines a return to 'impoverished bog . . . communities' similar in nature to relatively undamaged areas of peatland is a realistic goal. However, Anderson et al. (1997) also comment that current restoration techniques for fire damaged systems, based on seeding, whilst producing good vegetation cover and surface stabilization tend to produce

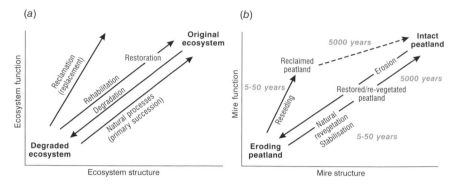

Figure 9.5 (a) Classification of restoration strategies. Reclamation is defined here as distinct from restoration effectively as a pragmatic approach to restoring system function without necessarily achieving near natural structure (after Dobson et al. 1997). (reprinted with permission from Dobson, A. P., Bradshaw, A. D. and Baker, A. J. M. 'Hopes for the future: Restoration ecology and conservation biology'. *Science* 277: 515–22. ©1997 AAAS). (b) Trajectories of change in restoration of eroded peatlands. Note the estimated timescales for the key transitions. At human timescales a restored and re-vegetated peatland is not transient but a long-term landscape component distinct from an uneroded peatland

broad expanses of very low species diversity unlike the natural mosaic of the surviving moorland. In essence this is reclaimed moorland (Figure 9.5a and 9.5b), *sensu* Dobson (1997), which may provide some of the functions of intact peatland but does not replicate the structure of the natural system (Table 9.3). In the conclusion to the Moorland Restoration Report/Anderson et al. (1997) state that suitable restoration approaches for areas of gully erosion do not exist. They question whether restoration is appropriate since some erosion is natural and the eroded peatlands of the Southern Pennines are effectively a 'type site' for erosion. Chapter 8 argued that erosion *and re-vegetation* are natural process in upland mires, and that the current extent of erosion is in some part due to suppression of natural re-vegetation. Under these circumstances it is suggested that some restoration of gullied sites is appropriate.

The gully blocking approaches discussed above provide a possible restoration approach, although the techniques are sufficiently new that long-term evidence of the efficacy of the approach is lacking. Some clarification of the aims of gully blocking in this context is required. Whereas the blocking of moorland drains is primarily focussed on raising water-tables

Table 9.3 Key functional changes associated with the main stable states identified in Figure 9.5b

Function	Intact	Restored/reclaimed	Eroding
DOC	Moderate	Moderate to high	High
POC	Low	Moderate	High
Productivity	High	High	Moderate
Agricultural use	Rough grazing and grouse moor	Restriction of grazing	Loss of productive area
Gaseous C exchange	CO_2 and CH_4 production	Increased CO_2	Increased CO_2
Water-table	High	Recovering	Reduced
Runoff	Rapid	Very rapid	Very rapid
Pollutant flux	Low particulate Moderate dissolved	Moderate particulate and dissolved	High particulate and dissolved

within degraded mires the effects of blocking gullies are more varied. Eroded gullies in upland peatland may exceed 3 metres in depth whereas typical drains are generally shallower. Naturally eroded systems are therefore not directly analogous to systems damaged by human intervention and even with large-scale blocking complete restoration of the water-table in the short term is not a realistic goal. Gully blocks are however analogous to the mechanism for gully re-vegetation through partial blocking and impedance of drainage as discussed in Chapter 8. Gully blocking may therefore promote re-vegetation of gully floors with the effect of limiting sediment export from the gully system and decreasing the degree of linkage between hillslope sediment sources and the stream channel (Chapter 4). The work on peatland sediment budgets reported in Chapters 4 and 8 suggests that to achieve these effects blocking and re-vegetation are required only at key points in the system, specifically the slope–channel interface. The evidence of naturally re-vegetated areas of gully erosion is that stabilization does not require a continuous vegetation cover but that more widespread re-vegetation can develop from these nodes as vegetation cover expands upstream and over a period of many tens of years onto the gully sides. Adopting this approach however involves much less intensive blocking than has typically been implemented and would be insufficient to significantly raise water-tables. It also leaves significant areas of the peatland bare and eroding (although not exporting sediment), which may conflict with ecological aims and be susceptible to local wind erosion.

The detailed approach to restoration through gully blocking is therefore dependent on the overall aims of the restoration project. True restoration (Figure 9.5a) is constrained by the mismatch between the circa 250 years it takes to erode a deep gully and the circa 5,000 years required to accumulate deep peat. A peatland restored through gully blocking is a physically stable but functionally altered system. It is argued that equivalence to the naturally re-vegetated state is an appropriate aim for gully restoration, preserving the distinctive eroded topography whilst limiting sediment and carbon flux from the system. The restored gully systems are water-gathering sites so that over the very long term enhanced peat growth may act to reduce topographic differentiation on the moor surface producing an approximation of intact mire. However the timescales necessary to achieve this are thousands of years so that this cannot be a primary restoration aim. Some caveats are required here as long-term assessments of gully-blocking techniques do not exist. There is the potential for long-term evolution of restored gully systems to an alternative stable state (e.g. cyclic cut and fill of the gully systems). This may be driven by either ecological processes (Suding et al. 2004), irreversible physical and chemical changes in the peat (Holden et al. 2004), or by hydrological and geomorphological processes associated with the eroded topography. Nevertheless if erosion and re-vegetation of upland peatlands is regarded as a natural part of peatland systems, then re-vegetated and/or restored gully systems represent a valuable component of a functioning peat landsystem.

9.5   Conclusions

9.5.1   The nature of upland peatlands

Peatlands are fundamentally hydrological systems. Water movements and energy flows through these systems drive the geomorphological and ecological processes which create and modify the landscape. Hydrology is central to understanding the geomorphology but the topographic structure of the peatland, and the geomorphological processes which create this, also provide important feedback to hydrological, biochemical and ecological processes. Chapter 1 presents a peat landsystems model (Figure 1.8) which integrates the geomorphological structures associated with mire instability into a conceptual model of an upland peatland. This model adopts a geomorphological focus emphasizing sources, sinks and flows of sediment (organic and mineral) which provides a structure for sediment budget investigations.

The major challenge in research in this area is to understand the peatland system in an integrated way. The premise of this book has been that the geomorphological component of this system has been understudied in relation to the other components of the system. The aim of this book has been to provide a systematic evaluation of the peatlands from a geomorphological perspective and as such provide a framework for future interdisciplinary approaches.

9.5.2   Geomorphological processes in upland peatlands

Fundamental to a sediment budget perspective on peatland geomorphology is consideration of sediment sources, sinks and the linkages between these. Sediment production on slopes is controlled by climatic parameters, particularly frost action and rainfall (Chapters 3 and 5). The sediment cascade dictates that loss from slopes is an important component of sediment supply lower down the geomorphic system in channel and on floodplains. However, not all sediment loss from slopes is transferred downslope. Deflation and oxidation take on special significance in peatland environments by virtue of the unique material properties of peat. Furthermore, at the small scale the processes of sediment supply are not fully understood. In particular the nature of within-storm sediment exhaustion requires further research, and the relative importance of deflation, oxidation and water erosion on bare peat surfaces needs to be evaluated.

The larger-scale patterns of fluvial erosion are well characterized using Bower's classification to define two end-members of a range of forms (Chapter 4). The range of potential forms between these end members is significant and the controls on the pattern of gully development require further investigation. The Bower scheme was essentially an erosional classification but increasingly we are recognizing deposition forms and re-vegetation of gully systems which are equally significant (Chapter 4). In particular Chapter 4 has demonstrated that the degree of re-vegetation is a key control on sediment flux. In systems where re-vegetation has significantly reduced the degree of slope–channel linkage the role of the perennial channel system as a sediment source is emphasized. Chapter 4 presents a conceptual model of sediment system behaviour in upland peatlands controlled by degree of erosion and re-vegetation.

On steeper ground mass movement may be a locally-important mechanism of peat erosion. The study of peat mass movements has the longest history of any area of peatland geomorphology but in the last 20 years a more systematic approach has been adopted in examining peat rapid mass movements (Chapter 5). There are two main types of failure – peat slides and bog bursts – distinguished on the basis of the peat characteristics where the landslides occur and the particular hydrological properties of different peat types. Hydrology is the critical controlling factor governing the mechanism of failure and spatial distribution of the mass movements. However, the range of peat mass movements needs to be defined more precisely and a better understanding of the geotechnical properties of peat on hillslopes is needed. Peat creep is also identified as a significant process on peat hillslopes but one that has been under-investigated and requires further work.

The conceptual model developed in Chapter 4 applies to systems dominated by fluvial action. Only in the last few years has quantification of the nature of wind erosion in upland landscapes begun to resolve a debate that has festered in the peatland geomorphic literature for nearly 50 years (Chapter 6). Recent measurements have conclusively demonstrated that both wind-driven rain and dry blow processes impose strong directional properties to the bulk sediment flux which produces a range of characteristic wind erosion and deposition landforms. However, such conclusions only apply to relatively exposed flat terrain; where slopes are present fluvial forms rapidly dominate (Chapter 6).

Understanding the upland peat landsystem as a unique set of geomorphic forms bound by a set of distinctive processes provides a tool for describing process–form relations and establishing linkages across different scales. Because of our improved understanding of the mechanisms and processes of peat erosion we can now make definite links between measured process rates and landform development. For example, process measurements of eroded peat flux under wind-driven rain can be directly related to the development of large areas of streamlined micro-relief hummocks and mounds (Chapter 6). Such associations were hitherto described only qualitatively. This improved understanding offers the opportunity to upscale from small-scale studies but also to downscale from predicted changes in forcing factors through identification of critical process linkages in the landsystem model.

9.5.3   The future of upland peatlands

Observations of natural re-vegetation imply that peat erosion is potentially a reversible process. In fact it may be that rates of re-vegetation are as important as rates of erosion in determining the extent of erosion within upland peatland landscapes. However, regional differences in rates of change and the trajectory of change exist for different peatland environments (Clement 2005). Predictions of widespread catastrophic erosion have not been proven correct but the degree of 'recovery' varies in space. A clearer understanding of controls on re-vegetation is central to understanding trajectories of peatland change in this context. Further research is required on the interaction of ecological and geomorphological processes to promote re-vegetation of eroded landscapes. Although significant re-vegetation of some eroded peatlands is observed this is not universal (Clement 2005). Indeed the palaeoecological evidence of increased erosion under conditions of moisture stress, and/or increased rainfall, suggest that further acceleration of erosion may be a consequence of predicted climate change. Clear process understanding in a sediment budget framework provides a sound basis for attempting to predict the complex response of peatland geomorphology to climate change.

9.5.4 Representativeness of the peatland system model

The empirical findings and theoretical ideas developed in this book are heavily based on the extensive body of work on the eroding peatlands of the UK. This volume represents the first systematic attempt to synthesize this knowledge within a geomorphological framework. As such it provides a foundation for a physically based geomorphological approach to conservation of these threatened upland landscapes.

Limited evidence from other geographical areas precludes an empirical assessment of the wider geographical applicability of these findings. The data on peatland hydrology and peat mass movements, where more international work has been done, suggest that, although there is some local variation relating to variable mire types, the physical forms and processes observed are similar due to generic properties of peat as a hydrological and geotechnical material. This implies that the findings of this volume, which are based on an understanding of the physical properties of peat and the erosional processes acting upon it, have the potential to be extrapolated to physically unstable peatlands in other parts of the world. Therefore, as Northern Hemisphere peatlands become increasingly stressed by climate change, permafrost melts, and surfaces become more unstable, the knowledge-base derived from the degraded peatlands of the UK will provide a useful starting point for understanding potentially more widespread future peatland erosion.

The motivation for writing this book came from a conviction that a general understanding of the geomorphology of upland peatlands lagged behind the understanding of their ecology and hydrology. The important research questions lie at the intersection of these three key peatland sciences, and real progress on the central scientific questions will require geomorphology to take its place alongside hydrology and ecology in interdisciplinary approaches to upland peat systems.

References

Acreman, M. (1991) The flood of July 25th on the Hermitage Water, Roxburgh-shire. *Scottish Geographical Magazine* 107(3): 170–8.

Åhman, R. (1976) The structure and morphology of minerogenic palsas in Northern Norway. *Biuletyn Peryglacjalny* 26: 25–31.

Aitkenhead, J. A., Hope, D. and Billett, M. F. (1999) The relationship between dissolved organic carbon in streamwater and soil organic carbon pools at different spatial scales. *Hydrological Processes* 13: 1289–302.

Aiton, W. (1811) Treatise on the origin, qualities, and cultivation of moss-earth, with directions for converting it into manure. Ayr, Scotland: Wilson and Paul.

Alexander, R. W., Coxon, P. and Thorn, R. H. (1986) A bog flow at Straduff Townland, County Sligo. *Proceedings of the Royal Irish Academy* 86B: 107–19.

Alexander, R. W., Coxon, P. and Tomlinson, R. W. (1985) Bog flows in south-east Sligo and South-west Leitrim. In *Irish Association for Quaternary Studies, Field Guide No. 8 for Sligo and West Leitrim*, edited by R. Thorn. Dublin: Irish Association for Quaternary Studies, pp. 58–80.

Anderson, E. W. (1977) Soil creep: An assessment of certain controlling factors with special reference to Upper Weardale, England. Unpublished PhD thesis. University of Durham, 585p.

Anderson, P. (1986) *Accidental moorland fires in the Peak District*. Bakewell: Peak Park Joint Planning Board.

Anderson, P. (1997) Fire damage on blanket mires. In *Blanket Mire Degradation: Causes, Consequences and Challenges*, edited by J. H. Tallis, R. Meade and P. D. Hulme. Macaulay Land Use Research Institute, Aberdeen, pp. 16–29.

Anderson, P. and Radford, E. (1994) Changes in vegetation following reduction in grazing pressure on the National Trust's Kinder Estate, Peak District, Derbyshire, England. *Biological Conservation* 69: 55–63.

Anderson, P., Tallis, J. and Yalden, D. (eds.) (1997) *Restoring moorland: Peak District moorland management project phase III report*. Bakewell: Peak District Moorland Management Project.

Apsimon, H. M., Kruse, M. and Bell, J. N. B. (1987) Ammonia emissions and their role in acid deposition. *Atmospheric Environment* 21(9): 1939–46.

Archer, D. (1992) *Land of singing waters: Rivers and great floods of Northumbria.* Spreddon Press: Northumbria, 217p.

Arden-Clarke, C. and Evans, R. (1993) Soil erosion and conservation in the United Kingdom. In *World soil erosion and conservation*, edited by D. Pimentel. Cambridge: Cambridge University Press, pp. 193–215.

Armstrong, A. (2005) Monitoring and modelling suspended sediment flux in British upland catchments. PhD thesis. University of Durham, 475p.

Armstrong, A. C. (1995) Hydrological model of peat-mound form with vertically varying hydraulic conductivity. *Earth Surface Processes and Landforms* 20(5): 473–7.

Arnalds, O. (2000) The Icelandic 'Rofabard' soil erosion features. *Earth Surface Processes and Landforms* 25: 17–28.

Ashmore, P., Brayshay, B. A., Edwards, K. J., Gilbertson, D. D., Grattan, J. P., Kent, M., Pratt, K. E. and Weaver, R. E. (2000) Allochthonous and autochthonous mire deposits, slope instability and palaeoenvironmental investigations in the Borve Valley, Barra, Outer Hebrides, Scotland. *Holocene* 10(1): 97–108.

Bagnold, R. A. (1941) *The physics of blown sand and desert dunes.* London: Methuen, 265p.

Bailey, A. (1879) A letter from Acting Governor Bailey to Governor Callaghan. *Quarterly Journal of the Geological Society of London* 35: 96–7.

Baird, A. J., Beckwith, C. W. and Heathewaite, A. L. (1997) Water movement in undamaged blanket peats. In *Blanket mire degradation: Causes, consequences and challenges*, edited by J. H. Tallis, R., Meade and P. Hulme. Aberdeen: British Ecological Society 128–39.

Baird, A. J., Price, J. S., Roulet, N. T. and Heathwaite, A. L. (2004) Special issue of *Hydrological Processes*: wetland hydrology and eco-hydrology. *Hydrological Processes* 18(2): 211–12.

Bakker, T. W. M. (1992) The shape of bogs from a hydrological point of view. *International Peat Journal* 4: 47–54.

Ball, D. F. (1964) Loss-on-ignition as an estimate of organic matter and organic carbon in non-calcareous soils. *Journal of Soil Science* 15: 84–92.

Ball, D. F. and Goodier, R. (1974) Ronas Hill, Shetland: A preliminary account of its ground pattern features resulting from the action of frost and wind. In *The Natural Environment of Shetland*, edited by R. Goodier. Edinburgh: Nature Conservancy Council, pp. 89–106.

Ballantyne, C. K. and Harris, C. (1994) *The periglaciation of Great Britain.* Cambridge: Cambridge University Press.

Ballantyne, C. K. and Whittington, G. W. (1987) Niveo-aeolian sand deposits on An Teallach, Wester Ross, Scotland. *Transactions of the Royal Society of Edinburgh: Earth Sciences* 78: 51–63.

Banas, K. and Gos, K. (2004) Effect of peat-bog reclamation on the physico-chemical characteristics of the ground water in peat. *Polish Journal of Ecology* 52(1): 69–74.

Barden, L. and Perry, P. L. (1968) Models of the consolidation process in peat soils. *Proceedings of the 3rd International Peat Congress, Quebec 1968*, pp. 119–27.

Barnes, F. A. (1963) Peat erosion in the Southern Pennines: Problems of inter-pretation. *East Midlands Geographer* 3: 216–22.

Beckwith, C. W. and Baird, A. J. (2001) Effect of biogenic gas bubbles on water flow through poorly decomposed blanket peat. *Water Resources Research* 37(3): 551–8.

Beckwith, C. W., Baird, A. J. and Heathwaite, A. L. (2003a) Anisotropy and depth-related heterogeneity of hydraulic conductivity in a bog peat. I: Labora-tory measurements. *Hydrological Processes* 17(1): 89–101.

Beckwith, C. W., Baird, A. J. and Heathwaite, A. L. (2003b) Anisotropy and depth-related heterogeneity of hydraulic conductivity in a bog peat. II: Modelling the effects on groundwater flow. *Hydrological Processes* 17(1): 103–13.

Begon, M., Harper, J. L. and Townsend, C. L. (1996) *Ecology: Individuals, popu-lations and communities.* Oxford: Blackwell.

Beilman, D. W. and Robinson, S. D. (2003) *Peatland permafrost thaw and landform type along a climatic gradient.* 8th International Conference on Permafrost, Zurich: Balkema Publishers, pp. 61–5.

Belyea, L. R. and Clymo, R. S. (1998) Do hollows control the rate of peat bog growth? In *Patterned mires and mire pools*, edited by V. Standen, J. Tallis and R. Meade. London: British Ecological Society, pp. 55–65.

Belyea, L. R. and Clymo, R. S. (2001) Feedback control of the rate of peat for-mation. *Proceedings of The Royal Society of London Series B-Biological Sciences* 268(1473): 1315–21.

Belyea, L. R. and Lancaster, J. (2002) Inferring landscape dynamics of bog pools from scaling relationships and spatial patterns. *Journal of Ecology* 90(2): 223–34.

Beven, K., Lawson, A. and McDonald, A. (1978) A landslip/debris flow in Bils-dale, North York Moors, September 1976. *Earth Surface Processes* 3: 407–19.

Birnie, R. V. (1993) Erosion rates on bare peat surfaces in Shetland. *Scottish Geographical Magazine* 109(1): 12–17.

Birse, E. L. (1980) Suggested amendments to the world soil classification to accommodate Scottish aeolian and mountain soils. *Journal of Soil Science* 31: 117–24.

Bishopp, D. W. and Mitchell, G. F. (1946) On a recent bog flow in Meenacharvy Townland, County Donegal. *Scientific Proceedings of the Royal Dublin Society* 24(17): 151–7.

Boardman, J. and Evans, R. (1994) Soil erosion in Britain: A review. In *Conserv-ing soil resources – European perspectives*, edited by R. J. Ricksen. Wallingford: CAB International, pp. 3–12.

Boatman, J. and Tomlinson, W. R. (1973) The Silver Flowe 1. Some structural and hydrological features of Brishie Bog and their bearing on pool formation. *Journal of Ecology* 61: 653–66.

Boelter, D. H. (1965) Hydraulic conductivity of peats. *Soil Science* 100(4): 227–31.

Boelter, D. H. (1968) Important physical properties of peat materials. *Proceedings of the 3rd International Peat Congress, Quebec 1968*, pp. 150–4.

Boelter, D. H. (1969) Physical properties of peats as related to degree of decomposition. *Soil Science Society of America Proceedings* 33(4): 606–9.

Bord na Móna (2001) The Peatlands of Ireland. http://www.bnm.ie/group/peat/peatlandsofireland.htm. Accessed December 2005.

Boudreau, S. and Rochefort, L. (1999) Etablissement de sphaignes reintroduites sous diverses communautes vegetales recolonisant les tourbieres apres l'exploitation. *Ecologie* 30: 53–62.

Bower, M. M. (1959) A summary of available evidence and a further investigation of the causes, methods and results of peat erosion in blanket peat. Unpublished MSc thesis. University of London.

Bower, M. M. (1960a) Peat erosion in the Pennines. *Advancement of Science* 64: 323–31.

Bower, M. M. (1960b) The erosion of blanket peat in the Southern Pennines. *East Midlands Geographer* 2(13): 22–33.

Bower, M. M. (1961) The distribution of erosion in blanket peat bogs in the Pennines. *Transactions of the Institute of British Geographers* (29): 17–30.

Bower, M. M. (1962) The cause of erosion in blanket peat bogs. *Scottish Geographical Magazine* 78: 33–43.

Bowes, D. R. (1960) A bog-burst in the Isle of Lewis. *Scottish Geographical Journal* 76: 21–3.

Bowler, M. and Bradshaw, R. H. W. (1985) Recent accumulation and erosion of blanket peat in the Wicklow Mountains, Ireland. *New Phytologist* 101: 543–50.

Bradshaw, R. and McGee, E. (1988) The extent and time-course of mountain blanket peat erosion in Ireland. *New Phytologist* 108(2): 219–24.

Bragg, O. M. (2002) Hydrology of peat-forming wetlands in Scotland. *Science of the Total Environment* 294(1–3): 111–29.

Bragg, O. M. and Tallis, J. H. (2001) The sensitivity of peat-covered upland landscapes. *Catena* 42(2–4): 345–60.

Bremner, J. M. and Jenkinson, D. S. (1960) Determination of organic carbon in soil. *Journal of Soil Science* 11: 394–402.

Bromley, J. and Robinson, M. (1995) Groundwater in raised mire systems: Models, mounds and myths. In *Hydrology and hydrochemistry of British wetlands*, edited by J. Hughes and A. L. Heathwaite. Chichester: Wiley, pp. 95–109.

Burt, T. P. (1992) The hydrology of headwater catchments. In *Rivers handbook*, edited by P. Calow and G. E. Petts. Oxford: Blackwell, pp. 3–28.

Burt, T. P. and Gardiner, A. T. (1984) Runoff and sediment production in a small peat covered catchment: Some preliminary results. In *Catchment experiments in fluvial geomorphology*, edited by T. P. Burt and D. E. Walling. Norwich: Geo Books, pp. 133–51.

Burt, T. P., Heathwaite, A. L. and Labadz, J. C. (1990) Runoff production in peat covered-catchment. In *Process studies in hillslope hydrology*, edited by M. G. Anderson and T. P. Burt. Chichester: Wiley, pp. 463–500.

Burt, T. P., Labadz, J. C. and Butcher, D. P. (1997) The hydrology and fluvial geomorphology of blanket peat: Implications for integrated catchment management. In *Blanket mire degradation: Causes, consequences and challenges*, edited by

J. H Tallis, R. Meade and P. Hulme. Aberdeen: British Ecological Society, pp. 121–7.

Burton, R. (1996) The peat resources of Great Britain (Scotland, England, Wales and Isle of Man). In *Global peat resources*, edited by E. Lappalainen. Jyskä (Finland): International Peat Society, pp. 79–86.

Caillier, M. and Visser, S. A. (1988) Observations on the dispersion and aggregation of clays by humic substances, short-term effects of humus-rich peat water on clay aggregation. *Geoderma* 43: 1–9.

Caine, N. (1980) The rainfall intensity–duration control of shallow landslides and debris flows. *Geografiska Annaler* 62A(1–2): 23–7.

Campbell, D. I. and Williamson, J. L. (1997) Evaporation from a raised peat bog. *Journal of Hydrology* 193(1–4): 142–60.

Campbell, D. R., Lavoie, C. and Rochefort, L. (2002) Wind erosion and surface stability in abandoned milled peatlands. *Canadian Journal of Soil Science* 82(1): 85–95.

Campbell, G. W. and Lee, D. S. (1996) Atmospheric deposition of sulphur and nitrogen species in the UK. *Freshwater Biology* 36(1): 151–67.

Campbell, I. B. (1981) Soil pattern of Campbell Island. *New Zealand Journal of Soil Science* 24: 111–35.

Campeau, S. and Rochefort, L. (1996) Sphagnum regeneration on bare peat surfaces: Field and greenhouse experiments. *Journal of Applied Ecology* 33(3): 599–608.

Campeau, S., Rochefort, L. and Price, J. S. (2004) On the use of shallow basins to restore cutover peatlands: Plant establishment. *Restoration Ecology* 12(4): 471–82.

Carey, S. K. and Woo, M. K. (2000) The role of soil pipes as a slope runoff mechanism, Subarctic Yukon, Canada. *Journal of Hydrology* 233(1–4): 206–22.

Carey, S. K. and Woo, M. K. (2002) Hydrogeomorphic relations among soil pipes, flow pathways, and soil detachments within a permafrost hillslope. *Physical Geography* 23(2): 95–114.

Carling, P. A. (1983) Particulate dynamics, dissolved and total load, in two small basins, Northern Pennines, UK. *Hydrological Sciences Journal-Journal Des Sciences Hydrologiques* 28(3): 355–75.

Carling, P. A. (1986a) Peat slides in Teesdale and Weardale, Northern Pennines, July 1983 – description and failure mechanisms. *Earth Surface Processes and Landforms* 11(2): 193–206.

Carling, P. A. (1986b) The Noon Hill flash floods; July 17th 1983: Hydrological and geomorphological aspects of a major formative event in an upland landscape. *Transactions of the Institute of British Geographers* NS 11: 105–18.

Carling, P. A., Glaister, M. S. and Flintham, T. P. (1997) The erodibility of upland soils and the design of pre-afforestation drainage networks in the United Kingdom. *Hydrological Processes* 11(15): 1963–80.

Carlsten, P. (1993) *Peat-geotechnical properties and up-to-date methods of design and construction*. State-of-the-art-report. 215. Linköping: Swedish Geotechnical Institute.

Caseldine, C. and Gearey, B. (2005) A multiproxy approach to reconstructing surface wetness changes and prehistoric bog bursts in a raised mire system at Derryville Bog, Co. Tipperary, Ireland. *The Holocene* 15(4): 585–601.

Caseldine, C., Thompson, G., Langdon, C. and Hendon, D. (2005) Evidence for an extreme climatic event on Achill Island, Co. Mayo, Ireland around 5200–5100 cal. yr bp. *Journal of Quaternary Science* 20(2): 169–78.

Caulfield, S. (1983) The Neolithic settlement of North Connaught. In *Landscape archaeology in Ireland*, edited by T. Reeves-Smith and F. Hammond. Oxford: British Archaeological Report, pp. 195–215.

Chappell, A. and Warren, A. (2003) Spatial scales of ^{137}Cs –derived soil flux by wind in a 25 km^2 arable area of eastern England. *Catena* 52(3): 209–34.

Charman, D. (2002) *Peatlands and environmental change*. Chichester: Wiley, 301p.

Chasar, L. S., Chanton, J. P., Glaser, P. H., Siegel, D. I. and Rivers, J. S. (2000) Radiocarbon and stable carbon isotopic evidence for transport and transformation of dissolved organic carbon, dissolved inorganic carbon, and CH_4 in a Northern Minnesota peatland. *Global Biogeochemical Cycles* 14(4): 1095–108.

Church, M. (2005) Continental drift. *Earth Surface Processes and Landforms* 30(1): 129–30.

Church, M. and Ryder, J. M. (1972) Paraglacial sedimentation; a consideration of fluvial processes conditioned by glaciation. *Bulletin of the Geological Society of America* 83(10): 3059–72.

Clark, J. M., Chapman, P. J., Adamson, J. K. and Lane, S. N. (2005) Influence of drought-induced acidification on the mobility of dissolved organic carbon in peat soils. *Global Change Biology* 11(5): 791–809.

Clement, S. (2005) The future stability of upland blanket peat following historical erosion and recent re-vegetation. Unpublished PhD thesis. Durham University, 387p.

Clymo, R. S. (1984) The limits to peat bog growth. *Royal Society Philosophical Transactions* B 303: 605–54.

Clymo, R. S. (2004) Hydraulic conductivity of peat at Ellergower Moss, Scotland. *Hydrological Processes* 18(2): 261–74.

Cole, K. L., Engstrom, D. R., Futyma, R. P. and Stottlemeyer, R. (1973) Past atmospheric deposition of metals in Northern Indiana measured in a peat core from Cowles Bog. *Nature* 245: 216–18.

Colhoun, E. A. (1965) The debris flow at Glendalough, Co. Wicklow and the bog-flow at Slieve Rushen, Co. Cavan, January 1965. *Irish Naturalist* 15: 199–206.

Colhoun, E. A., Common, R. and Cruickshank, M. M. (1965) Recent bog flows and debris slides in Northern Ireland. *Scientific Proceedings of the Royal Dublin Society* A 2: 163–74.

Conway, V. M. (1954) Stratigraphy and pollen analysis of Southern Pennine blanket peats. *Journal of Ecology* 42: 117–47.

Conway, V. M. and Millar, A. (1960) The hydrology of some small peat covered catchments in the Northern Pennines. *Journal of the Institute of Water Engineers* 14: 415–24.

Cooper, A. and Loftus, M. (1998) The application of multivariate land classification to vegetation survey in the Wicklow Mountains, Ireland. *Plant Ecology* 135:
229–41.

Cooper, A. and McCann, T. P. (1995) Machine peat cutting and land-use change
on blanket bog in Northern Ireland. *Journal of Environmental Management*
43(2): 153–70.

Cooper, A., McCann, T. P. and Hamill, B. (2001) Vegetation regeneration on
blanket mire after mechanized peat-cutting. *Global Ecology and Biogeography*
10(3): 275–89.

Couper, A., Immirzi, P. and Reid, E. (1997) The nature and extent of degradation in Scottish blanket mires. In *Blanket peat degradation: Causes, consequences,
challenges*, edited by J. H. Tallis, R. Meade and P. Hulme. British Ecological
Society Aberdeen.

Couper, P., Stott, T. and Maddock, I. (2002) Insights into river bank erosion
processes derived from analysis of negative erosion-pin recordings: Observations from three recent UK studies. *Earth Surface Processes and Landforms* 27(1):
59–79.

Couwenberg, J. (2005) A simulation model of mire patterning – revisited. *Eco-
graphy* 28(5): 653–61.

Crampton, C. B. (1911) *The vegetation of Caithness considered in relation to the
geology*. HM Geological Survey. Edinburgh: Turnbull and Spears, 132p.

Creighton, R. (2004) Landslides in Ireland. In *GeoHazards – Ireland at Risk*. The
Institute of Geologists of Ireland (IGI) Conference, Dublin 2004. Extended
Abstracts, 10–12.

Crisp, D. T. (1966) Input and output of minerals for an area of Pennine moorland
– importance of precipitation, drainage, peat erosion and animals. *Journal of
Applied Ecology* 3(2): 327–9.

Crisp, D. T., Rawes, M. and Welch, D. (1964) A Pennine peat slide. *Geographical
Journal* 130(4): 519–24.

Crofton, M. T. (1902) How Chat Moss broke out in 1526. *Transactions of the
Lancashire and Cheshire Antiquarian Society* 20: 139–44.

Crozier, M. J. (1986) *Landslides: Causes, consequences and environment*. London:
Croom Helm.

Cruickshank, M. M. and Tomlinson, R. W. (1988) *Northern Ireland peatland
survey*. Belfast Department of the Environment, Northern Ireland.

Cruickshank, M. M. and Tomlinson, R. W. (1990) Peatland in Northern Ireland:
Inventory and prospect. *Irish Geography* 23: 17–30.

Cummings, C. E. and Pollard, W. H. (1990) Cryogenetic characterization of peat
and mineral-cored palsas in Schefferville area, Quebec. *Collection Nordicana*
54: 95–102.

Daniels, S. (2002) The relationship between sub-catchment characteristics and
stream water chemistry: A study at Bleaklow, Derbyshire. Unpublished MSc
thesis, University of Manchester.

Dawson, J. J. C., Billett, M. F., Hope, D., Palmer, S. M. and Deacon, C. M.
(2004) Sources and sinks of aquatic carbon in a peatland stream continuum.
Biogeochemistry 70(1): 71–92.

Dawson, J. J. C., Billet, M. F., Neal, C. and Hill, S. (2002) A comparison of particulate, dissolved and gaseous carbon in two contrasting upland streams in the UK. *Journal of Hydrology* 257: 226–46.

De Lima, J. L. M. P. (1989) Raindrop splash anisotropy: Slope, wind and overland flow velocity effects. *Soil Technology* 2: 71–8.

De Lima, J. L. M. P., Van Dijk, P. M. and Spaan, W. P. (1992) Splash-saltation transport under wind-driven rain. *Soil Technology* 5: 151–66.

Delap, A. D., Farrington, A., Lloyd-Praeger, R. and Smyth, L. B. (1932) Report on a recent bog flow at Glencullin, County Mayo. *Scientific Proceedings of the Royal Dublin Society* 20(17): 181–92.

Delap, A. D. and Mitchell, G. F. (1939) On a recent bogflow in Powerscourt Mountain, Townland, Co. Wicklow. *Scientific Proceedings of the Royal Dublin Society* 22: 195–8.

Department of the Environment (1994) Peat slides. In *Landsliding in Great Britain*. HMSO. London.

Dietrich, W. E., Dunne, T., Humphrey, N. F. and Reid, L. M. (1982) Construction of sediment budgets for drainage basins. In *Sediment budgets and routing in forested drainage basins*, edited by F. J. Swanson, R. J. Janda, T. Dunne and D. N. Swanston. United States Department of Agriculture. General Technical Report PNW 141.

Dobson, A. P., Bradshaw, A. D. and Baker, A. J. M. (1997) Hope for the future: Restoration ecology and conservation biology. *Science* 277(5325): 515–22.

Dugan, D. (2004) North Mayo landslides. GeoHazards, The Institute of Geologists of Ireland (IGI) Conference, Dublin 2004. Extended Abstracts, 18–19.

Duncan, J. M. and Wright, S. G. (2005) *Soil strength and slope stability*. New Jersey: Wiley, 297p.

Dykes, A. P. and Kirk, K. J. (2000) Blanket bog failures. In *The geomorphology of the Cuilcagh Mountain, Ireland. A field guide for the British Geomorphological Research Group Spring Field Meeting*, May 2000, edited by J. Gunn, 26–35.

Dykes, A. P. and Kirk, K. J. (2001) Initiation of a multiple peat slide on Cuilcagh Mountain, Northern Ireland. *Earth Surface Processes and Landforms* 26(4): 395–408.

Dykes, A. P. and Kirk, K. J. (2006) Slope instability and mass movements in peat deposits. In *Peatlands: Evolution and records of environmental and climatic changes*, edited by I. P. Martini, A. Martinez Cortizas and W. Chesworth. Amsterdam: Elsevier.

Edil, T. B., Fox, P. J. and Lan, L. (1994) Stress-induced one-dimensional creep of peat. In *Advances in the understanding and modelling the mechanical behaviour of peat. Proceedings, Delft, Netherlands, 16–18 June, 1993*, edited by E. den Haan, R. Termaat and T. B. Edil. Rotterdam: A.A. Balkema, pp. 3–17.

Egglesmann, R., Heatwaite, A. L., Grosse-Brauckmann, G., Kuster, G. E., Naucke, W., Schuch, M. and Schweikle, V. (1993) Physical processes and properties of mires. In *Mires, process, exploitation and conservation*, edited by A. L. Heathwaite and K. Gottlich. Chichester: Wiley, pp. 171–262.

Ehrenfeld, J. G. (2000) Defining the limits of restoration: The need for realistic goals. *Restoration Ecology* 8: 2–9.

Elder, J. F., Rybicki, N. B., Carter, V. and Weintraub, V. (2000) Sources and yields of dissolved carbon in Northern Wisconsin stream catchments with differing amounts of peatland. *Wetlands* 20(1): 113–25.

Ellis, C. J. and Tallis, J. H. (2001) Climatic control of peat erosion in a North Wales blanket mire. *New Phytologist* 152: 313–24.

Erpul, G., Gabriels, D. and Janssens, D. (1998) Assessing the drop size distribution of simulated rainfall in a wind tunnel. *Soil Tillage and Research* 45: 455–63.

Erpul, G., Norton, L. D. and Gabriels, D. (2002a) Raindrop-induced and wind driven soil particle transport. *Catena* 47: 227–43.

Erpul, G., Norton, L. D. and Gabriels, D. (2002b) Erosion by wind-driven rain. In *Encyclopedia of Soil Science*, edited by R. Lal. New York: Dekker, pp. 285–93.

Erpul, G., Norton, L. D. and Gabriels, D. (2004) Splash-saltation trajectories of soil particles under wind-driven rain. *Catena* 47: 227–43.

Evans, M. G. and Burt, T. P. (1998) Contemporary erosion in the Rough Sike catchment, Moor House National Nature Reserve, North Pennines. In *Geomorphological studies in the North Pennines: Field guide*, edited by J. Warburton. Durham: BGRG: 37–45.

Evans, M. G., Burt, T. P., Clement, S. and Warburton, J. (2002) Erosion and re-vegetation of upland blanket peats: Implications for moorland management and carbon sequestration. Nature and People: Conservation and Management in the Mountains of Northern Europe, conference held at Pitlochry, Scotland, 7–9 November 2002. Unpublished poster.

Evans, M. G., Burt, T. P., Holden, J. and Adamson, J. K. (1999) Runoff generation and water table fluctuations in blanket peat: Evidence from UK data spanning the dry summer of 1995. *Journal of Hydrology* 221(3–4): 141–60.

Evans, M. G., Holden, J., Flitcroft, C. and Bonn, A. (2005) Understanding gully blocking in deep peat. *Moors for the Future Report* 4: 105.

Evans M. G. and Warburton, J. (2001) Transport and dispersal of organic debris (peat blocks) in upland fluvial systems. *Earth Surface Processes and Landforms* 26: 1087–102.

Evans, M. G. and Warburton, J. (2005) Sediment budget for an eroding peat-moorland catchment in Northern England. *Earth Surface Processes and Landforms* 30(5): 557–77.

Evans, M., Warburton, J. and Yang, J. (2006) Sediment budgets for eroding blanket peat catchments: Global and local implications of upland organic sediment budgets. *Geomorphology* 79(1–2) 45–57.

Farrell, C. A. and Doyle, G. J. (2003) Rehabilitation of industrial cutaway Atlantic blanket bog in County Mayo, north-nest Ireland. *Wetlands Ecology and Management* 11(1–2): 21–35.

Feehan, J. and O'Donovan, G. (1996) *The bogs of Ireland: An introduction to the natural, cultural and industrial heritage of Irish peatlands.* Dublin: University College Dublin, pp. 399–419.

Feldmeyer-Christe, E. (1995) La Vraconnaz, une tourbière en mouvement. Dynamique de la vegetation dans une tourbière soumise à un glissement de terrain. *Botanica Helvetica* 105: 55–73.

Feldmeyer-Christe, E. and Küchler, M. (2002) Onze ans de dynamique de la vegetation dans une tourbière soumise à un glissement de terrain. *Botanica Helvetica* 112: 103–20.

Feldmeyer-Christe, E. and Mulhauser, G. (1994) A moving mire – the burst bog of la Vraconnaz. *Fifth field symposium ICE International Mire conservation.* Birmensdorf: Swiss Federal Institute of Forest, Snow and Landscape Research, pp. 181–6.

Ferguson, P., Lee, J. A. and Bell, J. N. B. (1978) Effects of sulfur pollutants on growth of sphagnum species. *Environmental Pollution* 16(2): 151–62.

Fielding, A. H. and Howarth, P. F. (1999) *Upland habitats.* London: Routledge. 141p.

Fitzgibbon, J. E. (1981) Thawing of seasonally frozen ground in organic terrain in Central Saskatchewan. *Canadian Journal of Earth Sciences* 18: 1492–6.

Foster, D. R., Wright, H. E., Thelaus, M. and King, G. A. (1988) Bog development and landform dynamics in Central Sweden and South Eastern Labrador, Canada. *Journal of Ecology* 76(4): 1164–85.

Foulds, S. A. (2004) Wind-splash erosion and wind direction control on blanket peat geomorphology. Unpublished MSc thesis. University of Leeds, 80p.

Foulds, S. A. and Warburton, J. (2007a) Significance of wind-driven rain (wind-splash) in the erosion of blanket peat. *Geomorphology* 83: 183–92.

Foulds, S. A. and Warburton, J. (2007b) Wind erosion of blanket peat during a short period of surface desiccation (North Pennines, Northern England). *Earth Surface Processes and Landforms*, vol. 32, DOI: 10.1002/esp.1422.

Francis, I. S. (1987) Blanket peat erosion in mid-Wales: Two catchment studies. PhD thesis. University of Wales, Aberystwyth.

Francis, I. S. (1990). Blanket peat erosion in a mid-Wales catchment during two drought years. *Earth Surface Processes and Landforms* 15(5): 445–56.

Francis, I. S. and Taylor, J. A. (1989) The effect of forestry drainage operations on upland sediment yields: A study of two peat-covered catchments. *Earth Surface Processes and Landforms* 14(1): 73–83.

Fraser, C. J. D., Roulet, N. T. and Moore, T. R. (2001) Hydrology and dissolved organic carbon biogeochemistry in an ombrotrophic bog. *Hydrological Processes* 15(16): 3151–66.

Freeman, C., Evans, C. D., Monteith, D. T., Reynolds, B. and Fenner, N. (2001a) Export of organic carbon from peat soils. *Nature* 412: 785.

Freeman, C., Ostle, N. and Kang, H. (2001b) An enzymic 'latch' on a global carbon store – a shortage of oxygen locks up carbon in peatlands by restraining a single enzyme. *Nature* 409(6817): 149–9.

Gallart, F., Clotet-Perarnau, N., Bianciotto, O. and Puigdefabregas, J. (1994) Peat soil flows in Gahia del Buen Sucesco, Tierra del Fuego (Argentina). *Geomorphology* 9: 235–41.

Galvin, L. F. (1976) Physical properties of Irish peats. *Irish Journal of Agricultural Research* 15: 207–21.

Gardiner, A. T. (1983) Runoff and erosional processes in a peat-moorland catchment. Unpublished MPhil thesis. CNAA, Huddersfield Polytechnic.

Garnett, M. (1998) Carbon storage in Pennine moorland and response to climate change. Unpublished PhD thesis. University of Newcastle-Upon-Tyne, 302p.

Garnett, M. and Adamson, J. (1997) Blanket mire monitoring and research at Moor House National Nature Reserve. In *Blanket mire degradation causes, consequences and challenges*, edited by J. Tallis, R. Meade and P. Hulme. Aberdeen: British Ecological Society, pp. 116–17.

Garnett, M., Ineson, P. and Stevenson, A. C. (2000) Effects of burning and grazing on carbon sequestration in a Pennine blanket bog. *Holocene* 10(6): 729–36.

Garnett, M., Ineson, P., Stevenson, A. and Howard, D. (2001) Terrestrial organic carbon storage in a British moorland. *Global Change Biology* 7: 375–88.

Geerling, G. and Van Gestel, C. (1997) A study of erosion in Western Ireland. PhD thesis. Katholieke Universitiet Nijmegen.

Geikie, J. (1866) On the buried forests and peat deposits of Scotland and the changes in climate which they indicate. *Transactions of the Royal Society of Edinburgh* 24: 363–84.

Gerhardt, A. (1993) Review of impact of heavy metals on stream invertebrates with special emphasis on acid conditions. *Water Air and Soil Pollution* 66: 289–314.

Gerlach, T. (1967) Hillslope troughs for measuring sediment movement. *Revue Geomorphologie Dynamique* 4: 173.

Germann, P. F. (1990) Macropores and hydrologic hillslope processes. In *Process studies in hillslope hydrology*, edited by M. G. Anderson and T. P. Burt. Chichester: Wiley, pp. 327–63.

Gilman, K. and Newson, M. D. (1980) *Soil pipes and pipeflow: A hydrological study in upland Wales*. British Geomorphological Research Group Research Monograph No. 1. Norwich: Geo Books.

Girard, M., Lavoie, L. and Thériault, M. (2002) The regeneration of a highly disturbed ecosystem: A mined peatland in Southern Québec. *Ecosystems* 5(3): 274–88.

Glaser, P. H. (1998) The distribution and origin of mire pools. In *Patterned mires and mire pools*, edited by V. Standen, J. H. Tallis and R. Meade. London: British Ecological Society, pp. 4–25.

Glaser, P. H. and Janssens, J. A. (1986) Raised bogs in eastern North America – transitions in landforms and gross stratigraphy. *Canadian Journal of Botany-Revue Canadienne De Botanique* 64(2): 395–415.

Glaser, P. H., Siegel, D. I., Romanowicz, E. A. and Shen, Y. P. (1997) Regional linkages between raised bogs and the climate, groundwater, and landscape of north-western Minnesota. *Journal of Ecology* 85(1): 3–16.

Glatzel, S., Kalbitz, K., Dalva, M. and Moore, T. (2003) Dissolved organic matter properties and their relationship to carbon dioxide efflux from restored peat bogs. *Geoderma* 113(3–4): 397–411.

Glynn, T. E., Kirwan, R. W. and Wilson, N. E. (1968) Measurements of the dynamic response of peat subjected to repeated loading. *Proceedings of the 3rd International Peat Congress, Quebec 1968*, pp. 128–31.

Godwin, H. (1941) The factors which differentiate marsh, fen, bog and heath. *Chronical Botanical* 6(11).

Godwin, H. (1956) *The history of the British flora.* Cambridge: Cambridge University Press.

Gomez, B., Trustrum, N. A., Hicks, D. M., Rogers, K. M., Page, M. J. and Tate, K. R. (2003) Production, storage and output of particulate organic carbon: Waipaoa River basin, New Zealand. *Water Resources Research* 39(6): 1161.

Goodier, R. and Ball, D. F. (1975) Ward Hill, Hoy, Orkney: Patterned ground features and their origins. In *The Natural Environment of Orkney,* edited by R. Goodier. Edinburgh: Nature Conservancy Council, pp. 47–56.

Gore, A. J. P. (ed.) (1983) *Ecosystems of the world, 4a, mires: Swamp, bog, fen and moor.* Amsterdam: Elsevier.

Gore, A. J. P. and Godfrey, M. (1981) Reclamation of eroded peat in the Pennines. *Journal of Ecology* 69: 85–96.

Gorham, E. (1991) Northern peatlands: Role in the carbon cycle and probable responses to climatic warming. *Ecological Applications* 1: 182–95.

Gorham, E. and Rochefort, L. (2003) Peatland restoration: A brief assessment with special reference to sphagnum bogs. *Wetlands Ecology and Management* 11(1–2): 109–19.

Goudie, A. (ed.) (2004) *Encyclopedia of Geomorphology.* London: Routledge.

Grab, S. W. and Deschamps, C. L. (2004) Geomorphological and geoecological controls and processes following gully development in alpine mires, Lesotho. *Arctic Antarctic and Alpine Research* 36(1): 49–58.

Graham, J. (1984) Methods of stability analysis. In *Slope instability,* edited by D. Brunsden and D. B. Prior. Chichester: Wiley, pp. 171–214.

Graniero, P. A. and Price, J. S. (1999) The importance of topographic factors on the distribution of bog and heath in a Newfoundland blanket bog complex. *Catena* 36(3): 233–54.

Gregory, K. J., Gurnell, A. M. and Hill, C. T. (1985) The permanence of debris dams related to river channel processes. *Hydrological Sciences Journal-Journal Des Sciences Hydrologiques* 30(3): 371–81.

Grieve, I. C., Hipkin, J. A. and Davidson, D. A. (1994) Soil erosion sensitivity in upland Scotland. *Scottish Natural Heritage Research Survey and Monitoring Reports 24.* Edinburgh: Scottish Natural Heritage.

Griffiths, R. (1821) Report relative to the moving bog of Kilmaleady, in the King's County, made by the Order of the Royal Dublin Society. *Journal of the Royal Dublin Society* 141–4.

Groeneveld, E. V. G. and Rochefort, L. (2005) Polytrichum strictum as a solution to frost heaving in disturbed ecosystems: A case study with milled peatlands. *Restoration Ecology* 13(1): 74–82.

Grosvernier, P., Mathey, P. and A. Buttler (1995). Microclimate and physical properties of peat: New clues to the understanding of bog restoration processes. In *Restoration of temperate wetlands,* edited by B. D. Wheeler, S. C. Shaw, W. J. Fojt and R. A. Robertson. Chichester: Wiley, pp. 435–50.

Gunnarsson, U. and Rydin, H. (2000) Nitrogen fertilization reduces sphagnum production in bog communities. *New Phytologist* 147(3): 527–37.

Gurnell, A. M., Piegay, H., Swanson, F. J. and Gregory, S. V. (2002) Large wood and fluvial processes. *Freshwater Biology* 47(4): 601–19.

Gurney, S. D. (2001) Aspects of the genesis, geomorphology and terminology of palsas: Perennial cryogenic mounds. *Progress In Physical Geography* 25(2): 249–60.

Gustafson, E. J. (1998) Quantifying landscape spatial pattern: What is the state of the art? *Ecosystems* 1(2): 143–56.

Hagen, L. J., Wagner, L. E. and Skidmore, E. L. (1999) Analytical solutions and sensitivity analyses for sediment transport in WEPS. *Transactions of the ASAE* 42(6): 1715–22.

Haigh, M. J. (1977) The use of erosion pins in the study of slope evolution. In *Shorter Technical Methods*, edited by B. L. Finlayson. BGRG Technical Bulletin 18: 31–49.

Hall, D. J., Upton, S. L. and Marsland, G. W. (1994) Designs for a deposition gauge and a flux gauge for monitoring ambient dust. *Atmospheric Environment* 28(18): 2963–79.

Hall, G. and Cratchley, R. (2005) *Mechanisms of flooding in the Mawdach catchment.* Fourth Inter-Celtic Colloquium on Hydrology and Management of Water resources, Guimaraes, Portugal.

Hammond, R. (1979) *The peatlands of Ireland.* Soil Survey Bulletin 35. Dublin: An Foras Taluntais.

Hanrahan, E. T. (1954) An investigation of some physical properties of peat. *Geotechnique* 4: 108–23.

Harvey, A. M. (1994) Influence of slope/stream coupling on process interactions on eroding gully slopes: Howgill fells, northwest England. In *Process models and theoretical geomorphology*, edited by M. J. Kirkby. Chichester: Wiley, pp. 247–70.

Haycock, N. E., Trotter, S. and Hearn, K. (2004) Mapping and developing a strategic plan for the blocking of gullies for restoration of peat hydrology within the Dark Peak SSSI. River Restoration Centre Annual Conference. Durham: River Restoration Centre. http://www.therrc.co.uk/pdf/conferences/Summary%20of%20papers%202004.pdf. Accessed December 2005.

Heathwaite, A. L. (1993) Disappearing peat – regenerating peat? The impact of climate change on British peatlands. *Geographical Journal* 159(2): 203–8.

Heikurainen, L. (1963) On using ground water table fluctuations for measuring evapotranspiration. *Acta Forestalia Fennica* 76: 1–15.

Helenelund, K. V. and Hartikainen, J. (1972) In situ measurements of undrained shear strength of peat by helical auger tests. *Proceedings of the 4th International Peat Congress, Otaniemi, Finland, June 25–30:* Volume II – *Winning, harvesting, storage, transportation and processing of peat and sapropel for industrial, agricultural and horticultural purposes; geotechnics*, pp. 229–40.

Hemingway, J. E. and Sledge, W. A. (1945) A bog burst near Danby-in-Cleveland. *Proceedings of the Leeds Philosophical and Literature Society, Science* 4: 276–88.

Hemond, H. F. (1980) Biogeochemistry of Thoreau's Bog, Concord, Massachussetts. *Ecological Monographs* 50: 507–26.

Hemond, H. F. and Goldman, J. C. (1985) On non-darcian water-flow in peat. *Journal of Ecology* 73(2): 579–84.

Hendrick, E. (1990) A bog-flow at Bellacorrick Forest, County Mayo. *Irish Forestry* 47(1): 32–44.

Hewlett, J. D. and Hibbert, A. R. (1967) Factors affecting the response of small watersheds to precipitation in humid areas. In *Forest hydrology*, edited by W. E. Sopper and H. W. Lull. New York: Pergamon, pp. 275–90.

Higgitt, D. L., Warburton, J. and Evans, M. G. (2001) Sediment transfer in upland environments. In *Geomorphological processes and landscape change: Britain in the last 1000 years*, edited by D. L. Higgitt and E. M. Lee. Oxford: Blackwell, pp. 190–214.

Hoag, R. S. and Price, J. S. (1995) A field-scale, natural gradient solute transport experiment in peat at a Newfoundland blanket bog. *Journal of Hydrology* 172(1–4): 171–84.

Hoag, R. S. and Price, J. S. (1997) The effects of matrix diffusion on solute transport and retardation in undisturbed peat in laboratory columns. *Journal of Contaminant Hydrology* 28(3): 193–205.

Hobbs, N. (1985) Peat – morphology and properties. *Quarterly Journal of Engineering Geology* 18(3): 295.

Hobbs, N. B. (1986) Mire morphology and the properties and behaviour of some British and foreign peats. *Quarterly Journal of Engineering Geology* 19: 7–80.

Hogg, P., Squires, P. and Fitter, A. H. (1995) Acidification, nitrogen deposition and rapid vegetational change in a small valley mire in Yorkshire. *Biological Conservation* 71(2): 143–53.

Holden, J. (2001) Recent reduction of frost in the North Pennines. *Journal of Meteorology* 26(264): 369–74.

Holden, J. (2005) Controls of soil pipe frequency in upland blanket peat. *Journal of Geophysical Research-Earth Surface* 110(F1).

Holden, J. and Adamson, J. K. (2002) The Moor House long-term upland temperature record – new evidence of recent warming. *Weather* 57: 119–27.

Holden, J. and Burt, T. P. (2002a) Piping and pipeflow in a deep peat catchment. *Catena* 48: 163–99.

Holden, J. and Burt, T. P. (2002b) Laboratory experiments on drought and runoff in blanket peat. *European Journal of Soil Science* 53: 675–89.

Holden, J. and Burt, T. P. (2002c) Infiltration, runoff and sediment production in blanket peat catchments: Implications of field rainfall simulation experiments. *Hydrological Processes* 16(13): 2537–57.

Holden, J. and Burt, T. P. (2003a) Hydraulic conductivity in upland blanket peat: Measurement and variability. *Hydrological Processes* 17(6): 1227–37.

Holden, J. and Burt, T. P. (2003b) Hydrological studies on blanket peat: The significance of the acrotelm–catotelm model. *Journal of Ecology* 91(1): 86–102.

Holden, J. and Burt, T. P. (2003c) Runoff production in blanket peat covered catchments. *Water Resources Research* 39(7): art. no -.1191.

Holden, J., Burt T. P. and Cox, N. J. (2001) Macroporosity and infiltration in blanket peat: The implications of tension disc infiltrometer measurements. *Hydrological Processes* 15(2): 289–303.

Holden, J., Burt, T. P. and Vilas, M. (2002) Application of ground-penetrating radar to the identification of subsurface piping in blanket peat. *Earth Surface Processes and Landforms* 27(3): 235–49.

Holden, J., Chapman, P. J. and Labadz, J. C. (2004) Artificial drainage of peatlands: Hydrological and hydrochemical process and wetland restoration. *Progress in Physical Geography* 28(1): 95–123.

Holden, J., Evans, M. G., Burt, T. P. and Horton, M. (2006 In review) Impact of land drainage on peatland hydrology. *Journal of Environmental Quality*.

Holliday, V. J. (2003) Sediment budget for a North Pennine Upland Reservoir Catchment, UK. PhD thesis. University of Durham, 300p.

Honohane, J. (1697) Kapanhane bog flow. *Philosophical Transactions of the Royal Dublin Society*, Part XIX (Oct), 714–16.

Hope, D., Billet, M. F. and Cresser, M. S. (1994) A review of the export of carbon in river water: Fluxes and processes. *Environmental Pollution* 84: 301–24.

Hope, D., Billett, M. F. and Cresser, M. S. (1997a) Exports of organic carbon in two river systems in NE Scotland. *Journal of Hydrology* 193: 61–82.

Hope, D., Billet, M. F., Milne, R. and Brown, T. A. W. (1997b) Exports of organic carbon in British rivers. *Hydrological Processes* 11: 325–44.

Houghton, J., YDing, D., Griggs, M., Noguer, P., Linden, V. D. and Xiaosu, D. (eds.) (2001) *Climate change 2001: The scientific basis. Contribution of working group I to the third assessment report of the intergovernmental panel on climate change (IPCC).* IPCC assessment. Cambridge: Cambridge University Press.

Howard, P. J. A. and Howard, D. M. (1990) Use of organic carbon and loss-on-ignition to estimate soil organic matter in different soil types and horizons. *Biology and Fertility of Soils* 9: 306–10.

Huang, C. C. (2002) Holocene landscape development and human impact in the Connemara Uplands, Western Ireland. *Journal of Biogeography* 29(2) 153–65.

Hudleston, F. (1930) The cloudbursts of Stainmore, Westmorland, June 18th, 1930. *British Rainfall* 1930: 287–92.

Hughes, J. M. R. and Heathwaite, A. (1995a) Hydrology and hydrochemistry of British wetlands. In *Hydrology and hydrochemistry of British wetlands*, edited by J. Hughes and A. Heathwaite. Chichester: Wiley, pp. 1–8.

Hughes, J. M. R. and Heathwaite, A. (eds.) (1995b) *Hydrology and hydrochemistry of British wetlands.* Chichester: Wiley.

Hulme, M., Jenkins, G. J., Lu, X., Turnpenny, J. R., Mitchell, T. D., Jones, R. G., Lowe, J., Murphy, J. M., Hassell, D., Boorman, P., McDonald, R. and Hill, S. (2002) *Climate change scenarios for the United Kingdom: The UKCIP02 scientific report.* Norwich: Tyndall Centre for Climate Change Research, 120pp.

Hulme, P. D. (1986). The origin and development of wet hollows and pools on Craigeazle Mire, south-west Scotland. *International Peat Journal* 1: 15–28.

Hulme, P. D. and Blyth, A. W. (1985) Observations on the erosion of blanket peat in Yell, Shetland. *Geografiska Annaler Series A* 67: 119–22.

Hungr, O. and Evans, S. G. (1985) An example of a peat flow near Prince Rupert, British Columbia. *Canadian Geotechnical Journal* 22: 246–9.

Hungr, O., Morgan, G. C. and Kellerhals, R. (1984) Quantitative analysis of debris torrent hazards for design of remedial structures. *Canadian Geotechnical Journal* 21: 663–77.

Huscroft, C. A., Lipovsky, P. and Bond, J. D. (2004) Permafrost and landslide activity: Case studies from southwestern Yukon Territory. In *Yukon exploration and geology 2003*, edited by D. S. Emond and L. L. Lewis. Whitehorse: Yukon Geological Survey, pp. 107–19.

Hutchinson, J. N. (1980) The record of peat wastage in the East Anglian Fenlands at Holme Post, 1848–1978 AD. *Journal of Ecology* 68: 229–49.

Hutchinson, J. N. (1988) General report: Morphological and geotechnical parameters of landslides in relation to geology and hydrogeology. In *Proceedings 5th International Symposium on Landslides, Lausanne, Switzerland*, Vol. 1, 3–35, edited by C. Bonnard. Rotterdam: A.A. Balkema.

Hutchinson, S. M. (1995) Use of magnetic and radiometric measurements to investigate erosion and sedimentation in a British upland catchment. *Earth Surface Processes and Landforms* 20: 293–314.

Imeson, A. C. (1974) The origin of sediment in a moorland catchment with particular reference to the role of vegetation. In *Fluvial processes in instrumented catchments*, edited by K. J. Gregory and D. E. Walling. London: Institute of British Geographers, pp. 59–72.

Ingram, H. A. P. (1978) Soil layers in mires – function and terminology. *Journal of Soil Science* 29(2): 224–7.

Ingram, H. A. P. (1982) Size and shape in raised mire ecosystems – a geophysical model. *Nature* 297(5864): 300–3.

Ingram, H. A. P. (1983) Hydrology. In *Ecosystems of the world 4A. Mires: Swamp, bog, fen and moor*, edited by A. J. P. Gore. Amsterdam: Elsevier, pp. 67–158.

Ingram, H. A. P. (1987) Ecohydrology of Scottish peatlands. *Transactions of the Royal Society of Edinburgh Earth Sciences* 78: 287–96.

Ingram, H. A. P. and Bragg, O. M. (1984) *The diptotelmic mire: Some hydrological consequences reviewed*. Seventh International Peat Congress. Dublin: International Peat Society.

Ingram, H. A. P., Rycroft, D. W. and Williams, D. J. A. (1974) Anomalous transmission of water through certain peats. *Journal of Hydrology* 22: 213–18.

Ivanov, K. (1981) *Water movement in mirelands*. Translated from the Russian by A. Thompson and H. A. P. Ingram. London: Academic Press.

Jauhiainen, S., Laiho, R. and Vasander, H. (2002) Ecohydrological and vegetational changes in a restored bog and fen. *Annales Botanici Fennici* 39(3): 185–99.

Jennings, P. (2006) *Large-scale peat failures in Ireland*. GeoHazards – Ireland at Risk, The Institute of Geologists of Ireland (IGI) Conference, Dublin 2004.

Johnson, A. M. and Rodine, J. R. (1984) Debris flow. In *Slope instability*, edited by D. Brunsden and D. B. Prior. Chichester: Wiley, pp. 257–361.

Johnson, G. A. L. and Dunham, K. C. (1963) *The geology of Moor House*. Monographs of the Nature Conservancy Council Number Two. London: HMSO, 182p.

Johnson, R. M. and Warburton, J. (2002) Annual sediment budget of a UK mountain torrent. *Geografiska Annaler Series A-Physical Geography* 84A(2): 73–88.

Jones, J. A. A. (1981) *The nature of soil piping: A review of research.* British Geomorphological Research Group Monograph Series 3. Norwich: Geo Books.

Jones, J. A. A. (2004) Implications of natural soil piping for basin management in upland Britain. *Land Degradation & Development* 15(3): 325–49.

Jones, J. A. A. and Crane, F. G. (1984) Pipeflow and pipe erosion in the Maesnant experimental catchment. In *Catchment experiments in fluvial geomorphology*, edited by T. P. Burt and D. E. Walling. Norwich: Geobooks, pp. 55–70.

Jones, J. M. and Hao, J. (1993) Ombrotrophic peat as a medium for historical monitoring of heavy metal pollution. *Environmental Geochemistry and Health* 15(2–3): 67–74.

Kalembasa, S. J. and Jenkinson, D. S. (1973) A comparative study of titrimetric and gravimetric methods for the determination of organic carbon in soil. *Journal of Science, Food and Agriculture* 24: 1085–90.

Kellner, E. and Halldin, S. (2002) Water budget and surface-layer water storage in a sphagnum bog in Central Sweden. *Hydrological Processes* 16(1): 87–103.

Kennedy, G. W. and Price, J. S. (2005) A conceptual model of volume-change controls on the hydrology of cutover peats. *Journal of Hydrology* 302(1–4): 13–27.

Kinahan, G. H. (1897a) Bog slides and debacles. *Nature* 55: 268–9.

Kinahan, G. H. (1897b) Peat bog and debacles. *Transactions of the Institute of Civil Engineers of Ireland* 26: 98–123.

Kinako, P. D. S. and Gimingham, C. H. (1980) Heather burning and soil erosion on upland heaths in Scotland. *Journal of Environmental Management* 10(3): 277–84.

Kirk, K. J. (2001) Instability of blanket bog slopes on Cuilcagh Mountain, N.W. Ireland. Unpublished PhD thesis. University of Huddersfield.

Klinge, M. J. (1892) Über Mooraubrüche. *Jahrbucher Syst Pflanzenges Und Planzengeogr* 14: 426–61.

Klove, B. (1997) Settling of peat in sedimentation ponds. *Journal of Environmental Science and Health Part A – Environmental Science and Engineering & Toxic and Hazardous Substance Control* 32(5): 1507–23.

Klove, B. (1998) Erosion and sediment delivery from peat mines. *Soil and Tillage Research* 45(1–2): 199–216.

Kneale, P. E. (1987) Sensitivity of the groundwater mound model for predicting mire topography. *Nordic Hydrology* 18(4/5): 193–202.

Knighton, D. (1998) *Fluvial forms and processes.* London: Arnold.

Koelmans, A. A. and Prevo, L. (2003) Production of dissolved organic carbon in aquatic sediment suspensions. *Water Research* 37(9): 2217–22.

Kronvang, B. A. and Grant, L. A. R. (1997) Suspended sediment and particulate phosphorus transport and delivery pathways in an arable catchment, Gelbæk stream, Denmark. *Hydrological Processes* 11: 627–42.

Kurbatovdiv, J. (2002). Comparative ecosystem-atmosphere exchange of energy and mass in a European Russian and a Central Siberian bog. Interseasonal and interannual variability of energy and latent heat fluxes during the snowfree period. *Tellus* 54B: 497–513.

Labadz, J. (1988) Runoff and sediment production in blanket peat moorland: Studies in the Southern Pennines. Unpublished PhD thesis. Huddersfield Polytechnic.

Labadz, J. C., Burt, T. P. and Potter, A. W. R. (1991) Sediment yield and delivery in the blanket peat moorlands of the Southern Pennines. *Earth Surface Processes and Landforms* 16(3): 255–71.

Laine, A., Byrne, K. and Tuittilla, T. E. S. (2004) Vegetation composition and microform distribution in a blanket peatland. *Proceedings of the Irish Agricultural Research Forum*, Co. Offay.

Lal, R. (2003) Soil erosion and the global carbon budget. *Environment International* 29: 437–50.

Landva, A. O., Korpijaakko, E. O. and Pheeney, P. E. (1983) Geotechnical classification of peats and organic soils. In *Testing of peats and organic soils*, edited by A. P. M. Jarrett. Philadelphia: STM Special Technical Publication 820, pp. 37–51.

Landva, A. O. and Pheeney, P. E. (1980) Peat fabric and structure. *Canadian Geotechnical Journal* 17: 416–35.

Lapen, D. R., Price, J. S. and Gilbert, R. (2004) Modelling two-dimensional steady-state groundwater flow and flow sensitivity to boundary conditions in blanket peat complexes. *Hydrological Processes* 19(2): 371–86.

Large, A. R. G. (1991) The Slievenakilla bog-burst: Investigations into peat loss and recovery on an upland blanket bog. *Irish Naturalist Journal* 23(9): 354–9.

Large, A. R. G. and Hamilton, A. C. (1991) The distribution, extent and causes of peat loss in central and northwest Ireland. *Applied Geography* 11: 309–26.

Latimer, J. (1897) Some notes on the recent bog-slip in the Co. Kerry. *Transactions of the Institute of Civil Engineers Ireland* 26: 94–7.

Laudon, H., Kohler, S. and Buffam, I. (2004) Seasonal TOC export from seven boreal catchments in Northern Sweden. *Aquatic Sciences* 66(2): 223–30.

Lavoie, C., Grosvernier, P., Girard, M. and Marcoux, K. (2003) Spontaneous revegetation of mined peatlands: A useful restoration tool? *Wetlands Ecology and Management* 11(1–2): 97.

Lawler, D. M. (1986) River bank erosion and the influence of frost – a statistical examination. *Transactions of The Institute of British Geographers* 11(2): 227–42.

Lawler, D. M. (1991) A new technique for the automatic monitoring of erosion and deposition rates. *Water Resources Research* 27(8): 2125–8.

Lawler, D. M. (2005) The importance of high-resolution monitoring in erosion and deposition dynamics studies: Examples from estuarine and fluvial systems. *Geomorphology* 64(1–2): 1–23.

Lawler, D. M., West, J. R., Couperthwaite, J. S. and Mitchell, S. B. (2001) Application of a novel automatic erosion and deposition monitoring system at a channel bank site on the tidal river Trent, UK. *Estuarine Coastal And Shelf Science* 53(2): 237–47.

Lawlor, A. J. and Tipping, E. (2003) Metals in bulk deposition and surface waters at two upland locations in Northern England. *Environmental Pollution* 121(2): 153–67.

Ledger, D. C., Lovell, J. P. B. and McDonald, A. T. (1974) Sediment yield studies in upland catchment areas in south-east Scotland. *Journal of Applied Ecology* 11: 201–6.

Lee, J. A. and Tallis, J. H. (1973) Regional and historical aspects of lead pollution in Britain. *Nature* 245: 216–18.

Lee, J. A., Parsons, A. N. and Baxter, R. (1993) *Sphagnum* species and polluted environments, past and future. *Advances in Bryology* 5.

Legg, C. J., Maltby, E. and Proctor, M. C. F. (1992) The ecology of severe moorland fire on the North York Moors – seed distribution and seedling establishment of *calluna-vulgaris*. *Journal of Ecology* 80(4): 737–52.

Lewis, F. J. (1905) Plant remains in the Scottish peat mosses, part 1: The Scottish Southern Uplands. *Transactions of the Royal Society of Edinburgh* 41: 699–723.

Lewkowicz, A. G. (1992) Factors influencing the distribution and initiation of active-layer detachment slides on Ellesmere Island, Arctic Canada. In *Periglacial geomorphology*, edited by J. C. Dixon and A. D. Abrahams. Chichester: Wiley, pp. 223–50.

Limpens, J., Berendse, F. and Klees, H. (2004) How phosphorus availability affects the impact of nitrogen deposition on sphagnum and vascular plants in bogs. *Ecosystems* 7(8): 793–804.

Lindsay, R. (1995) *Bogs: The ecology, classification and conservation of obrotrophic mires*. Edinburgh: Scottish Natural Heritage, 119p.

Lindsay, R., Charman, D., Everingham, F., O'Reilly, R., Palmer, M., Rowell, T. and Stroud, D. (1988) *The flow country: The peatlands of Caithness and Sutherland*. Peterborough: NCC.

Lindsay, R. A., Riggall, J. and Burd, F. (1985) The use of small-scale surface patterns in the classification of British peatlands. *Aquilo Seria Botanica* 21: 69–79.

Livett, E. A., Lee, J. A. and Tallis, J. H. (1979) Lead, zinc and copper analyses of British blanket peats. *Journal of Ecology* 67: 865–91.

Livingstone, I. and Warren, A. (1996) *Aeolian geomorphology: An introduction*. Harlow: Longman.

Luoto, M. and Sepälä, M. (2000) Summit peats ('peat cakes') on the fells of Finnish Lapland: Continental fragments of blanket mires? *The Holocene* 10: 229–41.

Luukkainen, V. M. (1992) The flowing properties of peat slurry. In Proceedings of the 9th International Peat Congress, Uppsala, Sweden, 1992. Special Edition of the *International Peat Journal* 2(3): 137–45.

Mackay, A. W. (1993) The recent vegetational history of the Forest of Bowland, Lancashire. PhD thesis. Manchester.

Mackay, A. W. (1997) Peat erosion: Some implications for conservation and management of blanket mires and heather moorland in the UK. In *Conserving Peatlands*, edited by L. Parkyn, R. E. Stoneman and H. A. P. Ingram. Wallingford: CAB International, pp. 82–7.

Mackay, A. W. and Tallis, J. H. (1994) The recent vegetational history of the Forest of Bowland, Lancashire, UK. *New Phytologist* 128(3): 571–84.

Mackay, A. W. and Tallis, J. H. (1996) Summit-type blanket mire erosion in the Forest of Bowland, Lancashire, UK: Predisposing factors and implications for conservation. *Biological Conservation* 76: 31–44.

Mallik, A. U., Gimingham, C. H. and Rahman, A. A. (1984) Ecological effects of heather burning.1. Water infiltration, moisture retention and porosity of surface soil. *Journal of Ecology* 72(3): 767–76.

Maltby, E., Legg, C. J. and Proctor, M. C. F. (1990) The ecology of severe moorland fire on the North York Moors – effects of the 1976 fires, and subsequent surface and vegetation development. *Journal of Ecology* 78(2): 490–518.

Martinez Cortizas, A. and Garcia-Rodeja, E. (2002) Atmospheric Pb deposition in Spain during the last 4,600 years recorded by two ombrotrophic peat bogs and implications for the use of peat as archive. *The Science of the Total Environment* 292(1–2): 33–44.

Marutani, T., Kasai, M., Reid, L. M. and Trustrum, N. A. (1999) Influence of storm-related sediment storage on the sediment delivery from tributary catchments in the Upper Waipaoa River, New Zealand. *Earth Surface Processes and Landforms* 24(10): 881–96.

McCahon, C. P., Carling, P. A. and Pascoe, D. (1987) Chemical and ecological effects of a Pennine peat slide. *Environmental Pollution* 45: 275–89.

McDaniel, P. (2005) The twelve soil orders – soil taxonomy. Soil and Land Resources Division, University of Idaho, College of Agricultural and Life Sciences http://soils.ag.uidaho.edu/soilorders/. Accessed December 2005.

McEwen, L. J. and Withers, C. W. J. (1989) Historical records and geomorphological events: The 1771 'eruption' of Solway Moss. *Scottish Geographical Magazine* 105(3): 149–57.

McGreal, W. S. and Larmour, R. A. (1979) Blanket peat erosion: Theoretical considerations and observations from selected conservation sites in Slieveanorra Forest National Nature Reserve, Co. Antrim. *Irish Geography* 12: 57–67.

McHugh, M., Harrod, T. and Morgan, R. (2002) The extent of soil erosion in Upland England and Wales. *Earth Surface Processes and Landforms* 27(1): 99–107.

McKee, A. M. and Skeffington, M. (1997) Peat erosion and degradation in the Connemara Uplands. In *Blanket mire degradation: Causes, consequences and challenges*, edited by J. H. Tallis, R. Meade and P. D. Hulme. Aberdeen: British Ecological Society, pp. 81–2.

McMorrow, J. M., Cutler, M. E. J., Evans, M. G. and Al-Roichdi, A. (2004) Hyperspectral indices for characterizing upland peat composition. *International Journal of Remote Sensing* 25(2): 313–25.

Mills, A. J. (2002) Peat slides: Morphology, mechanisms and recovery. Unpublished PhD thesis. University of Durham.

Milne, R. and Brown, T. A. (1997) Carbon in the vegetation and soils of Great Britain. *Journal of Environmental Management* 49: 413–33.

Ministry of Agriculture Fisheries and Food (1993) *Code of good agricultural practice for the protection of soil*. London: MAFF Publications.

Mitchell, G. F. (1935) On a recent bog-flow in the County Clare. *Scientific Proceedings of the Royal Dublin Society* 21(27): 247–51.

Mitchell, G. F. (1938) On a recent bog flow in the County Wicklow. *Scientific Proceedings of the Royal Dublin Society* 22(4): 49–55.

Moeyersons, J. (1983) Measurements of splash-saltation fluxes under oblique rain. In *Rainfall simulation, runoff and soil erosion*, edited by J. Deploy. *Catena* Supplement 4: 19–31.

Mol, J. (1997) Fluvial response to Weichselian climate changes in the Niederlausitz (Germany). *Journal of Quaternary Science* 12(1): 41–60.

Moore, P. (1984) The classification of mires: An introduction. In *European mires*, edited by P. Moore. London: Academic Press, pp. 1–10.

Morgan, R. P. C. (1980) Soil erosion and conservation in Britain. *Progress in Physical Geography* 4: 24–47.

Moseley, M. P. (1972) Gully systems in blanket peat, Bleaklow, North Derbyshire. *East Midlands Geographer* 5(5): 235–44.

Moss, C. E. (1913) *The vegetation of the Peak District*. Cambridge: Cambridge University Press.

Mulvaney, J. (1879) Avalanche of peat in the Falkland Islands. *Proceedings of the Royal Geographical Society and Monthly Record of Geography. New Monthly Series* 1(12): 803–5.

Muschamp Perry, J. J. (1893) The waterspout (or cloud burst) on the Cheviots. *British Rainfall* 1893: 14–17.

Mylona, S. 1993. *Trends of sulphur dioxide emissions, air concentrations and depositions of sulphur in Europe since 1880*. EMEP/MSC-W Report 2/93. Oslo: The Norwegian Meteorological Institute.

Myślińska, E. (2003) Classification of organic soils for engineering geology. *Geological Quarterly* 47(1): 39–42.

Naden, P. S. and McDonald, A. T. (1989) Statistical modeling of water color in the uplands – the Upper Nidd catchment 1979–1987. *Environmental Pollution* 60(1–2): 141–63.

Naiman, R. J. (1982) Characteristics of sediment and organic carbon export from pristine boreal forest watersheds. *Canadian Journal of Fish and Aquatic Science* 39: 1699–718.

Nakamura, F. and Swanson, F. J. (1993) Effects of coarse woody debris on morphology and sediment storage of a mountain stream system in Western Oregon. *Earth Surface Processes and Landforms* 18(1): 43–61.

Nel, W., Holness, S. and Meiklejohn, K. I. (2003) Observations on rapid mass movements and screes on Sub-Antarctic Marion Island. *South African Journal of Science* 99: 177–81.

Newall, A. M. and Hughes, J. M. R. (1995). Microflow environments of aquatic plants in flowing water wetlands. In *Hydrology and hydrochemistry of British wetlands*, edited by J. M. R. Hughes and A. L. Heathwaite. Chichester: Wiley.

Nichol, D. and Farmer, I. W. (1998) Settlement over peat of the A5 at Pant Dedwydd near Cerrigydrundion, North Wales. *Engineering Geology* 50: 299–307.

Nickling, W. G. (1988) The initiation of particle movement by wind. *Sedimentology* 35: 499–511.

O'Connell, C. (2002) Irish peatland conservation council blanket bogs information sheet. http://www.ipcc.ie/infoblanketbogfs.html. Accessed March 2005.

Ogden, J. (1976) Nicholson unpublished poems. *The Bradford Antiquary: The Journal of the Bradford Historical and Antiquarian Society* 49: 37–44.

Oke, T. R. (1987) *Boundary layer climates*. London: Methuen.

Oksanen, P. O., Kuhry, P. and Alekseeva, R. N. (2001) Holocene development of the Rogovaya River Peat Plateau, European Russian Arctic. *Holocene* 11(1): 25–40.

Ouseley, R. (1788) An account of the moving of a bog, and the formation of a lake, in the County of Galway, Ireland. *Transactions of the Royal Irish Academy* 2: 3–6.

Outcalt, S. I. (1971) Algorithm for needle ice growth. *Water Resources Research* 7(2): 394–7.

Pattinson, V. A., Butcher, D. P. and Labadz, J. C. (1994). The management of water color in peatland catchments. *Journal of the Institution of Water and Environmental Management* 8(3): 298–307.

Peak District National Park Authority (2001) *Biodiversity action plan for the Peak District National Park*. Bakewell: Peak District National Park Authority.

Pearce-Higgins, J. W. and Yalden, D. W. (1997) The effect of resurfacing the Pennine Way on recreational use of blanket bog in the Peak District National Park. *Biological Conservation* 82: 337–43.

Pearsall, W. (1950) *Mountains and moorlands*. London: Fontana.

Pearsall, W. (1956) Two blanket bogs in Sutherland. *Journal of Ecology* 44: 493–516.

Pedersen, H. S. and Hasholt, B. (1995) Influence of wind speed on rain splash erosion. *Catena* 24: 39–54.

Pemberton, M. (2001) Soils in the Lake Pedder area. In *Lake Pedder: Values and restoration*, edited by C. Sharples. Hobart: University of Tasmania, pp. 61–6.

Philips, J., Tallis, J. and Yalden, D. (eds.) (1981) *Peak District moorland erosion study: Phase 1 report*. Bakewell: Peak Park Joint Planning Board.

Phillips, J. D. (1989) Fluvial sediment storage in wetlands. *Water Resources Bulletin* 25(4): 867–73.

Phillips, M. E. (1954) Studies in the quantitative morphology and ecology of *Eriophorum-angustifolium roth. II.* Competition and dispersion. *Journal of Ecology* 42(1): 187–210.

Pitkanen, A., Turunen, J. and Tolonen, K. (1999) The role of fire in the carbon dynamics of a mire, Eastern Finland. *Holocene* 9(4): 453–62.

Pollard, E. and Miller, A. (1968) Wind erosion in the East Anglian Fens. *Weather* 23: 415–17.

Praeger, R. L. (1897a) Bog-bursts, with special reference to the recent disaster in Co. Kerry. *The Irish Naturalist* 6: 141–62.

Praeger, R. L. (1897b) A bog-burst seven years after. *The Irish Naturalist* 6: 201–3.

Praeger, R. L. (1906) The Ballycumber bog slide. *The Irish Naturalist* 7: 1–6.

Press, M. C., Woodin, S. J. and Lee, J. A. (1986) The potential importance of an increased atmospheric nitrogen supply to the growth of ombrotrophic sphagnum species. *New Phytologist* 103(1): 45–55.

Price, J. S. (1991) Evaporation from a blanket bog in a foggy coastal environment. *Boundary-Layer Meteorology* 57(4): 391–406.

Price, J. S. (2003) Role and character of seasonal peat soil deformation on the hydrology of undisturbed and cutover peatlands. *Water Resources Research* 39(9): art. no.-1241.

Price, J. S., Rochefort, L. and Campeau, S. (2002) Use of shallow basins to restore cutover peatlands: Hydrology. *Restoration Ecology* 10(2): 259–66.

Price, J. S., Rochefort, L. and Quinty, F. (1998) Energy and moisture considerations on cutover peatlands: Surface microtopography, mulch cover and sphagnum regeneration. *Ecological Engineering* 10(4): 293–312.

Price, J. S. and Schlotzhauer, S. M. (1999) Importance of shrinkage and compression in determining water storage changes in peat: The case of a mined peatland. *Hydrological Processes* 13: 2591–601.

Price, J. S. and Whitehead, G. S. (2001) Developing hydrologic thresholds for sphagnum recolonisation on an abandoned cutover bog. *Wetlands Ecology and Management* 21(1): 32–40.

Pye, K. and Paine, A. D. M. (1983) Nature and source of aeolian deposits near the summit of Ben Arkle, northwest Scotland. *Geologie en Mijnbouw* 6313–18.

Radforth, N. W. (1962) Organic terrain and geomorphology. *Canadian Geographer* VI(3–4): 166–71.

Radley, J. (1962) Peat erosion on the high moors of Derbyshire and West Yorkshire. *East Midlands Geographer* 3(17): 40–50.

Radley, J. (1965) Significance of major moorland fires. *Nature* 205: 1254–9.

Rastall, R. H. and Smith, B. (1906) Tarns on the Haystacks Mountain, Buttermere, Cumberland. *Geological Magazine*, Decade V, Vol. III: 406–12.

Ratcliffe, D. (1977) *A nature conservation review: The selection of biological sites of national importance.* Cambridge: Cambridge University Press.

Reeve, A. S., Siegel, D. I. and Glaser, P. H. (2000) Simulating vertical flow in large peatlands. *Journal of Hydrology* 227(1–4): 207–17.

Reeve, A. S., Warzocha, J., Glaser, P. H. and Siegel, D. I. (2001) Regional ground-water flow modeling of the glacial lake Agassiz Peatlands, Minnesota. *Journal of Hydrology* 243(1–2): 91–100.

Renberg, I., Brannval, M., Bindler, R. and Emteryd, O. (2000) Atmospheric lead pollution history during four millennia (2000 BC to 2000 AD) in Sweden. *Ambio* 29(3): 150–6.

Reynolds, W. D., Brown, D. A., Mathur, S. P. and Overend, R. P. (1992) Effect of in situ gas accumulation on the hydraulic conductivity of peat. *Soil Science* 153(5): 397–408.

Rhodes, N. and Stevenson, T. (1997) Palaeoenvironmental evidence for the importance of fire as a cause of erosion of British and Irish blanket peats. In *Blanket mire degradation: Causes, consequences and challenges*, edited by J. Tallis, R. Meade and P. Hulme. Aberdeen: British Ecological Society, pp. 64–78.

Richards, J. R. A., Wheeler, B. D. and Willis, A. J. (1995) The growth and value of *Eriophorum angustifolium* Honck in relation to the revegetation of eroding blanket peat. In *Restoration of temperate wetlands*, edited by B. D. Wheeler, S. C. Shaw, W. J. Fojt and R. A. Robertson. Chichester: Wiley.

Robert, E. C., Rochefort, L. and Garneau, M. (1999) Natural revegetation of two block-cut mined peatlands in Eastern Canada. *Canadian Journal of Botany-Revue Canadienne De Botanique* 77(3): 447–59.

Robinson, M. and Blyth, K. (1982) The effect of forest drainage operations on upland sediment yields: A case study. *Earth Surface Processes and Landforms* 7: 85–90.

Rochefort, L. (2000) *Sphagnum* – a keystone genus in habitat restoration. *The Bryologist* 103(3): 503–8.

Rochefort, L. and Bastien, D. F. (1998) Reintroduction of *Sphagnum* into harvested peatlands: Evaluation of various methods for protection against dessication. *Ecoscience* 5(1): 117–27.

Rochefort, L., Quinty, F., Campeau, S., Johnson, K. and Malterer, T. (2003) North American approach to the restoration of *Sphagnum* dominated peatlands. *Wetlands Ecology and Management* 11(1–2): 3–20.

Rogers, N. W. and Selby, M. J. (1980) Mechanisms of shallow translational landsliding during summer rainstorm: North Island, New Zealand. *Geografiska Annaler* 62A(1–2): 11–21.

Romanov, V. V. (1968) *Hydrophysics of bogs*. Translated by N. Kaner and edited by A. Heimann. Jerusalem: Israel Progress in Science Translations, 299p.

Rothwell, J. J., Evans, M., Lindsay, J. and Allott, T. E. A. (2007) Scale-dependant spatial variability in peatland lead pollution in the Southern Pennines, UK. *Environmental Pollution* 145: 111–20.

Rothwell, J. J., Robinson, S. G., Evans, M. G., Yang, J. and Allott, T. E. H. (2005) Heavy metal release by peat erosion in the Peak District, Southern Pennines, UK. *Hydrological Processes* 19(15): 2973–89.

Rudberg, S. (1968) Wind erosion – the preparation of maps showing the direction of eroding winds. *Biuletyn Peryglacjalny* 17: 181–93.

Rycroft, D. W., Williams, D. J. A. and Ingram, H. A. P. (1975) The transmission of water through peat 1. Review. 2. Field experiments. *Journal of Ecology* 63(2): 535–68.

Salonen, V. and Laaksonen, M. (1994) Effects of fertilisation, liming, watering, and tillage on plant colonisation of bare peat surfaces. *Annales Botanici Fennici* 31(1): 29–36.

Samuelsson, G. (1910) Scottish peat mosses. A contribution to the knowledge of the late-quaternary vegetation and climate of north-western Europe. *Bulletin of the Geological Institute of Upsala* 10: 197–260.

Schlotzhauer, S. M. and Price, J. S. (1999) Soil water flow dynamics in a managed cutover peat field, Quebec: Field and laboratory investigations. *Water Resources Research* 35(12): 3675–83.

Schumm, S. A. (1979) Geomorphic thresholds: The concept and its applications. *Transactions of the Institute of British Geographers* 4: 485–515.

Selby, M. J. (1993) *Hillslope materials and processes* (second edn.). Oxford: Oxford University Press, 451p.

Selkirk, J. M. (1996) Peat slides on subantarctic Macquarie Island. *Zeitschrifte für Geomorphologie*, Supplementband 105: 61–72.

Selkirk, J. M. and Saffigna, L. J. (1999) Wind and water erosion of a peat and sand area on subantarctic Macquarie Island. *Arctic Antarctic and Alpine Research* 31(4): 412–20.

Seppälä, M. (2001) Strong deflation on palsas in Finnish Lapland. *Transactions of the Japanese Geomorphological Union* 22(4): C–216.

Seppälä, M. (2004) *Wind as a geomorphic agent in cold climates*. Cambridge: Cambridge University Press.

Shaw, S. C., Wheeler, B. D., Kirby, P., Phillipson, P. and Edmunds, R. (1996) Literature review of the historical effects of burning and grazing of blanket bog and upland wet heath. *English Nature research reports 172*. Peterborough: English Nature.

Shimwell, D. W. (1974) Sheep grazing intensity in Edale, Derbyshire, 1692–1747, and its effect on blanket peat erosion. *Derbyshire Archaeological Journal* 94: 35–40.

Shotyk, W. (1988) Review of the inorganic geochemistry of peats and peatland waters. *Earth Science Reviews* 25: 95–176.

Shotyk, W. (2002) The chronology of anthropogenic, atmospheric Pb deposition recorded by peat cores in three minerogenic peat deposits from Switzerland. *The Science of the Total Environment* 292(1–2): 19–31.

Shotyk, W., Norton, S. A. and Farmer, J. G. (1997) Summary of the workshop on peat bog archives of atmospheric metal deposition. *Water Air and Soil Pollution* 100(3–4): 213–19.

Shotyk, W., Weiss, D., Appleby, P. G., Cheburkin, A. K., Frei, R., Gloor, M., Kramers, J. D., Reese, S. and Van Der Knaap, W. O. (1998) History of atmospheric lead deposition since 12,370 ^{14}C yr BP from a peat bog, Jura Mountains, Switzerland. *Science* 281(5383): 1635–40.

Shroder, J. F. Jr. (1976) Mass movement on Nyika Plateau, Malawi. *Zeitschrift für Geomorphologie*, 20(1): 56–77.

Sirvent, J., Desir, G., Gutierrez, M., Sancho, C. and Benito, G. (1997) Erosion rates in badland areas recorded by collectors, erosion pins and profilometer techniques (Ebro Basin, NE Spain). *Geomorphology* 18(2): 61–75.

Skeffington, R., Wilson, E., Maltby, E., Immirzi, P. and Putwain, P. (1997) Acid deposition and blanket mire degradation and restoration. In *Blanket mire degradation: Causes, consequences and challenges*, edited by J. H. Tallis, R. Meade and P. D. Hulme. Aberdeen: Macaulay Land Use Research Institute, pp. 29–37.

Slattery, M. C., Gares, P. A. and Phillips, J. A. (2002) Slope-channel linkage and sediment delivery on North Carolina coastal plain cropland. *Earth Surface Processes and Landforms* 27(13): 1377–87.

Smith, S. V., Renwick, W. H., Buddmeier, R. W. and Crossland, C. J. (2001) Budgets of soil erosion and deposition for sediments and sedimentary organic carbon across the conterminous United States. *Global Biogeochemical Cycles* 15 (3): 697–707.

Söderblom, R. (1974) Organic matter in Swedish clays and its importance for quick clay formation. *Proceedings of the Swedish Geotechnical Institute* 26. 89p.

Sollas, W. J., Lloyd Praeger, R., Dixon, A. F. and Delap, A. (1987) Report of the committee consisting of Professor W. J. Sollass, LLD, FRS., R. Lloyd Praeger, BA, BE, A. F. Dixon, MB and A. Delap, BA, BE, appointed by the Royal Dublin Society to investigate the recent Bog-flow in Kerry. *Scientific Proceedings of the Royal Dublin Society*, VIII, V: 475–508.

Soulsby, C., Rodgers, P., Smart, R., Dawson, J. and Dunn, S. (2003) A tracer-based assessment of hydrological pathways at different spatial scales in a mesoscale Scottish catchment. *Hydrological Processes* 17(4): 759–77.

Standen, R. (1897) Bog bursts. *The Irish Naturalist* 1 (6): 224.

Statham, I. (1977) *Earth surface sediment transport.* Oxford: Clarendon Press.

Stevenson, A., Jones, V. and Battarbee, R. (1992) A palaeoecological evaluation of peat erosion. In *Peatland ecosystems and man – an impact assessment*, edited by O. Bragg, P. Hulme, H. Ingram and R. Robertson. Dundee: Dundee University, Dept. of Biological Sciences, pp. 363–6.

Stewart, A. J. and Lance, A. N. (1991) Effects of moor drains on the hydrology and vegetation of North Pennines blanket bog. *Journal of Applied Ecology* 28: 1105–17.

Stoneman, R. and Brooks, S. (eds.) (1997) *Conserving bogs. The management handbook.* Edinburgh: The Stationery Office.

Stott, T. (1997) A comparison of stream bank erosion processes on forested and moorland streams in the Balquhidder Catchments, Central Scotland. *Earth Surface Processes and Landforms* 22(4): 383–99.

Stott, T. A., Ferguson, R. I., Johnson, R. C. and Newson, M. D. (1986) Sediment budgets in forested and unforested basins in upland Scotland. In *Drainage basin sediment delivery, proceedings of the Albuquerque symposium*, edited by R. F. Hadley. Albuquerque: IAHS Publication 159, pp. 57–68.

Strack, M., Kellner, E. and Waddington, J. M. (2005) Dynamics of biogenic gas bubbles in peat and their effects on peatland biogeochemistry. *Global Biogeochemical Cycles* 19(1): art. no.-GB1003.

Suding, K. N., Gross, K. L. and Houseman, G. R. (2004) Alternative states and positive feedbacks in restoration ecology. *Trends In Ecology & Evolution* 19(1): 46–53.

Sundberg, S. and Rydin, H. (2002) Habitat requirements for establishment of sphagnum from spores. *Journal of Ecology* 90(2): 268–78.

Sutton, M. A., Asman, W. A. H., Ellermann, T., Van Jaarsveld, J. A., Acker, K., Aneja, V., Duyzer, J., Horvath, L., Paramonov, S., Mitosinkova, M., Tang, Y. S., Achermann, B., Gauger, T., Bartniki, J., Neftel, A. and Erisman, J. W. (2003) Establishing the link between ammonia emission control and measurements of reduced nitrogen concentrations and deposition. *Environmental Monitoring and Assessment* 82(2): 149–85.

Svensson, G. (1988) Fossil plant-communities and regeneration patterns on a raised bog in south Sweden. *Journal of Ecology* 76(1): 41–59.

Swanson, D. K. and Grigal, D. F. (1988) A simulation-model of mire patterning. *Oikos* 53(3): 309–14.

Swanson, F. J., Janda, R. J., Dunne, T. and Swanston, D. N. (eds.) (1982) *Sediment budgets and routing in forested drainage basins*. USDA General Technical Report PNW – 141.

Sykes, J. M. and Lane, A. M. J. (1996) *United Kingdom environmental change network: Protocols for standard measurements at terrestrial sites*. London: Natural Environment Research Council.

Talling, P. J. and Sowter, M. J. (1999) Drainage density on progressively tilted surfaces with different gradients, Wheeler Ridge, California. *Earth Surface Processes and Landforms* 24(9): 809–24.

Tallis, J. H. (1964a) Studies on Southern Pennine peats II. The pattern of erosion. *Journal of Ecology* 52: 333–44.

Tallis, J. H. (1964b) Studies on Southern Pennine peats III. The behaviour of *Sphagnum*. *Journal of Ecology* 52(2): 345–53.

Tallis, J. H. (1965) Studies on Southern Pennine peats IV. Evidence of recent erosion. *Journal of Ecology* 53: 509–20.

Tallis, J. H. (1969) The blanket bog vegetation of the Berwyn Mountains, North Wales. *Journal of Ecology* 57: 509–20.

Tallis, J. H. (1973) Studies on Southern Pennine peats V. Direct observations on peat erosion and peat hydrology at Featherbed Moss, Derbyshire. *Journal of Ecology* 61(1): 1–22.

Tallis, J. H. (1985a) Mass movement and erosion of a Southern Pennine blanket peat. *Journal of Ecology* 73: 283–315.

Tallis, J. H. (1985b) Erosion of blanket peat in the Southern Pennines: New light on an old problem. In *The geomorphology of north west England*, edited R. H. Johnson. Manchester: Manchester University Press, pp. 113–32.

Tallis, J. H. (1987) Fire and flood at Holme Moss: Erosion processes in an upland blanket mire. *Journal of Ecology* 75: 1099–129.

Tallis, J. H. (1994) Pool-and-hummock patterning in a Southern Pennine blanket mire II. The formation and erosion of the pool system. *Journal of Ecology* 82(4): 789–803.

Tallis, J. H. (1995) Climate and erosion signals in British blanket peats: The significance of *Racomitrium lanuginosum* remains. *Journal of Ecology* 83(6): 1021–30.

Tallis, J. H. (1997a) Peat erosion in the Pennines: The badlands of Britain. *Biologist* 44: 277–9.

Tallis, J. H. (1997b) The Southern Pennine experience: An overview of blanket mire degradation. In *Blanket mire degradation: Causes, consequences and challenges*, edited by J. H. Tallis, R. Meade and P. D. Hulme. Aberdeen: Macaulay Land Use Research Institute, pp. 7–16.

Tallis, J. H. (1997c) The pollen record of *Empetrum nigrum* in Southern Pennine peats: Implications for erosion and climate change. *Journal of Ecology* 85(4): 455–65.

Tallis, J. H. (1998) Growth and degradation of British and Irish blanket mires. *Environmental Review* 6: 81–122.

Tallis, J. H. (2001) Bog bursts. *Biologist* 48: 218.

Tallis, J. H. and Livett, E. A. (1994) Pool-and-hummock patterning in a Southern Pennine blanket mire I. Stratigraphic profiles for the last 2,800 years. *Journal of Ecology* 82(4): 775–88.

Tallis, J. H., Meade, R. and Hulme, P. D. (1997) *Blanket mire degradation: Causes, consequences and challenges.* Proceedings. Mires Research Group. Aberdeen: Macaulay Land Use Research Institute, 222p.

Tallis, J. H. and Switsur, V. R. (1973) Studies on Southern Pennine peats VI. Radiocarbon-dated pollen diagram from Featherbed Moss, Derbyshire. *Journal of Ecology* 61(3): 743–51.

Tallis, J. H. and Yalden, D. (1983) *Peak District moorland restoration project phase II report: Re-vegetation trials.* Bakewell: Peak Park Joint Planning Board.

Tansley, A. (1939) *The British Isles and their vegetation.* Cambridge: Cambridge University Press.

Terry, J. P. and Shakesby, R. A. (1993) Soil hydrophobicity effects on rainsplash: Simulated rainfall and photographic evidence. *Earth Surface Processes and Landforms* 18: 519–25.

Thorp, M. and Glanville, P. (2003) Mid-Holocene sub-blanket-peat alluvia and sediment sources in the Upper Liffey Valley, Co. Wicklow, Ireland. *Earth Surface Processes and Landforms* 28(9): 1013–24.

Tipping, E., Smith, E. J., Lawlor, A. J., Hughes, S. and Stevens, P. A. (2003) Predicting the release of metals from ombrotrophic peat due to drought-induced acidification. *Environmental Pollution* 123(2): 239–53.

Tobin, P. J. & Co. (2003) *Report on the landslides at Dooncarton, Glengad, Brnachuille and Pollathomais, County Mayo.* Unpublished technical report, 39p.

Tomlinson, R. W. (1979) Water levels in peatlands and some implications for runoff and erosional processes. In *Geographical approaches to fluvial processes,* edited by A. F. Pitty. Norwich: Geo Books, pp. 149–62.

Tomlinson, R. W. (1981a) A preliminary note on the bog-burst at Carromaculla, Co. Fermanagh, November 1979. *Irish Naturalist Journal* 20(8): 313–16.

Tomlinson, R. W. (1981b) The erosion of peat in the uplands of Northern Ireland. *Irish Geography* 14: 51–64.

Tomlinson, R. W. and Gardiner, T. (1982) Seven bog slides in the Slieve-an-Orra Hills, County Antrim. *Journal of Earth Science Royal Dublin Society* 5: 1–9.

Tranter, D. (1999) Case study: Wingecarribee Swamp, water wet or dry? *Proceedings of the Water and Wetlands Management Conference, November 1998.* New South Wales: Nature Conservation Council of NSW, pp. 90–7.

Trimble, S. W. (1981) Changes in sediment storage in the Coon Creek Basin, Driftless Area, Wisconsin, 1853 to 1975. *Science* 214(4517): 181–3.

Troll, C. (1944) Strukturböden, Solifluktion und Frostklimate der Erde. *Geologische Rundschau* 34: 545–694.

Tufnell, L. (1969) The range of periglacial phenomena in Northern England. *Biuletyn Peryglacjalny* 19: 292–323.

Tuittila, E. S., Komulainen, V. M., Vasander, H. and Laine, J. (1999) Restored cut-away peatland as a sink for atmospheric CO_2. *Oecologia* 120(4): 563–74.

Tuittila, E. S., Rita, H., Vasander, H. and Laine, J. (2000) Vegetation patterns around *Eriophorum vaginatum* l. Tussocks in a cut-away peatland in Southern

Finland. *Canadian Journal of Botany-Revue Canadienne de Botanique* 78(1): 47–58.

Tuittila, E. S., Vasander, H. and Laine, J. (2003) Success of re-introduced *sphagnum* in a cutaway peatland. *Boreal Environment Research* 8(3): 245–50.

Turunen, C. and Turunen, J. (2003) Development history and carbon accumulation of a slope bog in oceanic British Columbia, Canada. *Holocene* 13(2): 225–38.

UKCIP (2005) Climate impacts, United Kingdom Climate Change Impacts Programme. http://www.ukcip.org.uk. Accessed December 2005.

Van Seters, T. E. and Price, J. S. (2001) The impact of peat harvesting and natural regeneration on the water balance of an abandoned cutover bog, Quebec. *Hydrological Processes* 15(2): 233–48.

Van Seters, T. E. and Price, J. S. (2002) Towards a conceptual model of hydrological change on an abandoned cutover bog, Quebec. *Hydrological Processes* 16(10): 1965–81.

Vardy, S. R., Aitken, A. E. and Bell, T. (2000) *Mid-Holocene palaeoenvironmental history of Easter Axel Heiberg Island: Evidence from a rapidly accumulating peat deposit.* Geo-Canada 2000: The Millennium Geoscience Summit. Calgary: Canadian Society of Exploration Physicists.

Vasander, H., Tuittila, E. S., Lode, E., Lundin, L., Ilomets, M., Sallantaus, T., Heikkilä, R., Pitkänen, M. L. and Laine, J. (2003) Status and restoration of peatlands in Northern Europe. *Wetlands Ecology and Management* 11(1–2): 51–63.

Victoria National Park Association (2005) Should cattle return to fire affected areas. http://www.cowpaddock.com/return.html. Accessed June 2005.

Vidal, H. (1966) Die moorbruchkatastrophe bei Schönberg/Oberbayern am 13./14.6.1960. *Zeit Deutsche Geologische Gesellschaft* 115(2/3): 770–82.

Vieira, G., Mora, C. and Gouveia, M. M. (2004) Oblique rainfall and contemporary geomorphological dynamics. *Hydrological Processes* 18: 807–24.

Vile, M. A., Wieder, R. K. and Novak, M. (2000) 200 years of Pb deposition throughout the Czech Republic; patterns and sources. *Environmental Science and Technology* 34(1): 12–21.

von Post, L. (1924) *Das genetische System der organogenen Bildungen Schwedens.* Comité International de Pédologie IV Commission 22.

Waddington, J. M. and Roulet, N. T. (1997) Groundwater flow and dissolved carbon movement in a boreal peatland. *Journal of Hydrology* 191(1–4): 122–38.

Waine, J., Brown, J. M. B. and Ingram, H. A. P. (1985) Non-darcian transmission of water in certain humified peats. *Journal of Hydrology* 82(3–4): 327–39.

Walkley, A. and Black, I. A. (1934) An examination of the Degtjareff method for determining soil organic matter and a proposed modification of the chromic acid titration method. *Soil Science* 37: 29–38.

Walling, D. E. (1983) The sediment delivery problem. *Journal of Hydrology* 65(1–3): 209–37.

Walling, D. E., Owens, P. N. and Leeks, G. J. L. (1999) Rates of contemporary overbank sedimentation and sediment storage on the floodplains of the main channel systems of the Yorkshire Ouse and River Tweed, UK. *Hydrological Processes* 13(7): 993–1009.

Walling, D. E. and Webb, B. W. (1981) Water quality. In *British rivers*, edited by J. Lewis. London: Allen and Unwin, pp. 126–69.

Walling, D. E. and Webb, B. W. (1987) Suspended load in gravel-bed rivers: UK experience. In *Sediment transport in gravel-bed rivers*, edited by C. R. Thorne, J. C. Bathurst and R. D. Key. Hoboken: Wiley, pp. 691–723.

Walsh, S. J., Butler, D. R. and Malanson, G. P. (1998) An overview of scale, pattern, process relationships in geomorphology: A remote sensing and GIS perspective. *Geomorphology* 21(3–4): 183–205.

Warburton, J. (2003) Wind splash erosion of bare peat on UK upland moorlands. *Catena* 52: 191–207.

Warburton, J., Evans, M. G. and Johnson, R. M. (2003) Discussion on 'the extent of soil erosion in upland England and Wales'. *Earth Surface Processes and Landforms* 28(2): 219–23.

Warburton, J., Higgitt, D. L. and Mills, A. J. (2003) Anatomy of a Pennine peat slide, Northern England. *Earth Surface Processes and Landforms* 28: 457–73.

Warburton, J., Holden, J. and Mills, A. J. (2004) Hydrological controls of surficial mass movements in peat. *Earth-Science Reviews* 67(1–2): 139–56.

Ward, W. H. (1948) A slip in a flood defence bank constructed on a peat bog. *Geotechnique* 5: 154–63.

Webb, B. W. and Walling, D. E. (1984) Magnitude and frequency characteristics of suspended sediment transport in Devon rivers. In *Catchment experiments in fluvial geomorphology*, edited by T. P. Burt and D. E. Walling. Norwich: Geo Books, pp. 399–415.

Weiss, D., Shotyk, W., Boyle, E. A., Kramers, J. D., Appleby, P. G. and Cheburkin, A. K. (2002) Comparative study of the temporal evolution of atmospheric lead deposition in Scotland and eastern Canada using blanket peat bogs. *The Science of the Total Environment* 292(1–2): 7–18.

Weiss, D., Shotyk, W. and Kempf, O. (1999) Archives of atmospheric lead pollution. *Naturwissenschaften* 86: 262–75.

Werritty, A. and Ingram, H. A. P. (1985) Blanket mire erosion in the Scottish Borders in July 1983. *British Ecological Society Bulletin* 16: 202–3.

Wheeler, B. and Proctor, M. (2000) Ecological gradients, subdivisions and terminology of north-west European mires. *Journal of Ecology* 88: 187–203.

Wheeler, B. D. (1995) Introduction: Restoration and wetlands. In *Restoration of temperate wetlands*, edited by B. D. Wheeler, S. C. Shaw, W. Fojt and R. A. Robertson. Chichester: Wiley, pp. 1–18.

Wheeler, B. D., Shaw, S. C., Fojt, W. J. and Robertson, R.A. (eds.) (1995) *Restoration of temperate wetlands*. Chichester: Wiley.

White, J. (1930) Recolonisation after peat cutting. *Proceedings of the Royal Irish Academy, Section B: Biological, Geological and Chemical Science* 39: 453–76.

White, P., Labadz, J. and Butcher, D. (1996) Sediment yield estimates from reservoir studies: An appraisal of variability in the Southern Pennines of the UK.

In *Erosion and sediment yield: Global and regional perspectives*, edited by D. Walling and B. Webb. IAHS publication 236, pp. 163–73.

Wilford, G. E. (1965) A peat landslide in Sarawak, Malaysia, and its significance in relation to washouts in coal seams. *Journal of Sedimentary Petrology* 36(1): 244–7.

Wilson, N. E. (1972) Cell structure of organic soils under shearing stresses. *Proceedings of the 4th International Peat Congress, Otaniemi, Finland, June 25–30; Volume II – Winning, harvesting, storage, transportation and processing of peat and sapropel for industrial, agricultural and horticultural purposes; geotechnics*, pp. 291–303.

Wilson, P., Griffiths, D. and Carter, C. (1996) Characteristics, impacts and causes of the Carntogher bog-flow, Sperrin Mountains, Northern Ireland. *Scottish Geographical Magazine* 112(1): 39–46.

Wilson, P. and Hegarty, C. (1993) Morphology and causes of recent peat slides on Skerry Hill, County Antrim, Northern Ireland. *Earth Surface Processes and Landforms* 18: 593–601.

Wilson, S. J. and Cooke, R. U. (1980) Wind erosion. In *Soil erosion*, edited by M. J. Kirkby and R. P. Morgan. Chichester: Wiley, pp. 217–51.

Winch, S., Ridal, J. and Lean, D. (2002) Increased metal bioavailability following alteration of freshwater dissolved organic carbon by ultraviolet b radiation exposure. *Environmental Toxicology* 17(3): 267–74.

Wishart, D. and Warburton, J. (2001) An assessment of blanket mire degradation and peatland gully development in the Cheviot Hills, Northumberland. *Scottish Geographical Journal* 117(3): 185–206.

Worrall, F., Burt, T. P. and Adamson, J. (2003) Controls on the chemistry of runoff from an upland peat catchment. *Hydrological Processes* 17(10): 2063–83.

Worrall, F., Burt, T. P. and Adamson, J. (2004) Can climate change explain increases in DOC flux from upland peat catchments? *Science of the Total Environment* 326(1–3): 95–112.

Worrall, F., Burt, T. P., Jaeburn, R. Y., Shedden, R. and Warburton, J. (2002) Release of dissolved organic carbon from upland peat. *Hydrological Processes* 16(17): 3487–504.

Worrall, F., Burt, T. P. and Shedden, R. (2003) Long-term records of riverine dissolved organic matter. *Biogeochemistry* 64(2): 165–78.

Worrall F., Reed, M., Warburton, J. and Burt, T. P. (2003) Carbon budget for British upland peat catchment. *Science of the Total Environment* 312: 133–46.

Yang, J. (2005) Monitoring and modelling sediment flux from a blanket peat catchment in the Southern Pennines. Unpublished PhD thesis. School of Geography, University of Manchester.

Yeloff, D. E., Labadz, J. C., Hunt, C. O., Higgitt, D. L. and Foster, I. D. L. (2005) Blanket peat erosion and sediment yield in an upland reservoir catchment in the Southern Pennines, UK. *Earth Surface Processes and Landforms* 30(6): 717–33.

Yeo, M. (1997) Blanket mire degradation in Wales. In *Blanket mire degradation: Causes, consequences and challenges,* edited by J. H. Tallis, R. Meade and P. D. Hulme. Aberdeen: Macaulay Land Use Research Institute, pp. 101–15.

Zuidhoff, F. S. (2002) Recent decay of a single palsa in relation to weather conditions between 1996 and 2000 in Laivadalen, Northern Sweden. *Geografiska Annaler* 84A(2): 103–11.

Name Index

Note: page numbers in italics denote figures, tables or illustrations

Subject Index

Note: page numbers in italics denote figures, tables or illustrations